48

W9-DBQ-094

JAMES L. McCARTNEY, M. D.
223 STEWART AVENUE
GARDEN CITY, N. Y.

THE Living Brain

THE
Living Brain

W. GREY WALTER

New York · W·W· NORTON & COMPANY · INC · 1953

PRINTED IN THE UNITED STATES OF AMERICA
BY THE HADDON CRAFTSMEN, INC., SCRANTON, PA.

1

Contents

art of touch · Refinement of the blind · Why deafness is so hard to
bear · Some processes of vision · Why the eye must scan the scene ·
Its million transmission cables · Projection room of the brain.

Illustrations

9

Figure

Foreword

THIS BOOK is intended for general reading, for those who are interested in themselves and other creatures. Most of it is matter of recent discovery, but it has been presented simply so that even the most unexpected information should be intelligible to all. To some, the subject itself may seem risky: after giving a series of talks on it in the Home Service of the BBC I was told that one or two listeners said they felt a kind of impudicity about brain surveying brain, as if suddenly coming upon themselves for the first time naked in a looking-glass. Our peeping here is as innocent as Alice and kinder than Analysis. While mirroring indeed some parts of the human organism too long hidden, this book will be found a gentle book. There is no immodest exposure, no baring of the soul; nor any shattering of illusions, except perhaps for those who may have been so simple-minded as to think the mechanism of mind simple. Neither the new facts of life nor their philosophical consequences are presented as conclusive. Fresh realms of knowledge and conjecture are explored without pretence of reaching the limits of either, their boundaries always receding as we advance. Thus it is also a pious book if measured by standards of habitual reverence for the known as well as the unknown in face of successive revelations. Omar said: "I have learnt nothing from life except my own amazement at it." That is too little, but it is a beginning. It is an aspect of the matter which to some readers—modest, thoughtful, religious—may seem more important than the excitement of discoveries, whatever their clinical or social utility, or the desirability of removing error from our way of thinking about the brain. And in this reverent attitude to Man the author is with them, standing bareheaded among the villagers where "still they gaz'd and still the wonder grew That one small head could carry all he knew."

W.G.W.

THE Living Brain

". . . . an enchanted loom where millions of flashing shuttles weave a dissolving pattern, always a meaningful pattern though never an abiding one; . . ."

Sir Charles Sherrington, O.M.

CHAPTER 1

Lords of the Earth

Consider the original of all things, the matter they are made of, the alterations they must run through, and the result of the change. And that all this does no manner of harm.

Marcus Aurelius

By BRAIN is meant, in the first instance, something more than the pinko-grey jelly of the anatomist. It is, even to a scientist, the organ of imagination. "Enchanted loom" it was called by a great physiologist. Another has likened it to a calm lake on which ripple-systems weave patterns. The first image is a reminder that magic may be a function of mechanism. The second invites us to embark on the surface of something deeper than we know, and subject to storms.

With licence of such teaching, we may begin by saying that *cogito ergo sum* is physiologically true. Man, for our present purpose, is specifically what he is by virtue of thought, and owes his survival in the struggle for existence to the development of that supreme function of brain. He is *sapiens*, the thinking species of genus *Homo*—the discerning, discreet and judicious one, even if he does not always live up to all these meanings of the name he has given himself. But the brain has many other functions, so many and so various that it may be well to proceed first by elimination.

No other animal is equipped for being *sapiens*. It is in fact a difference of equipment and not of opportunity. In terms of behaviour, the gist of it is that, when we come across something new, we do not necessarily respond to it at once in a particular manner. We think it over. We can imagine making one of a number of possible responses, and imagine it so clearly that we can see whether it would be, if we made it, a mistake, without having to commit ourselves to action. We can make our errors in a thought and reject them in another thought, leaving no trace of error in us.

Very early in the human story the brain must have acquired the mechanism of what we recognise in action as imagination, calculation, prediction. Later came the processes of abstract reason and the control of what we call violence. The operation of these mental controls, as will be seen, can be recorded as electrical eddies swirling in subtle patterns through the brain. But our most sensitive instruments, amplifying the electrical changes ten million times or more, detect only isolated and intermittent elements of these higher functions in the brains of other animals.

Thus the mechanisms of the brain reveal a deep physiological division between man and ape, deeper than the superficial physical differences of most distant origin. If the title of soul be given to the higher functions in question, it must be admitted that the other animals have only a glimmer of the light that so shines before men. Aristotle's frontier of learning stands. The nearest creature to us, the chimpanzee, cannot retain an image long enough to reflect on it, however clever it may be in learning tricks or getting food that is placed beyond its natural reach. Unable to rehearse the possible consequences of different responses to a stimulus, without any faculty of planning, the apes and other animals cannot learn to control their feelings, first step towards independence of environment and eventual control of it. The activity of the animal brain is not checked to allow time for the choice of one among several possible responses, but only for the one reflex or conditioned response to emerge. The monkey's brain is still in thrall to its senses. *Sentio ergo sum* might be the first reflection of a slightly inebriated ape, as it is often the last of alcoholic man; so near and yet so far apart, even then, are they.

The brain of lion, tiger, rhinoceros and other powerful animals also lacks the mechanism of imagination, or we should not be here to discuss the matter. They cannot envisage changes in their environment, so they have never sought to alter it in all their efforts to retain lordship of their habitat. In case of flood, fire, frost or desiccation they can only seek to replace their lost amenities by ranging the forest, scanning desert or mountain, until they find them elsewhere, or die. It was not power that won them survival in the earth's changes of climate, but thermostasis. Fortunately for them—and for us—the great reptiles never acquired that knack of keeping the body temperature from changing with the environment. Nor have their degenerate descendants.

> No shiver greets the wintry blast,
> No drop of sweat the sun.

The insects found a way out of the dilemma without fur or feathers, without shiver or sweat. Their challenge to human supremacy would be vastly more serious than it is, had they not adopted a method of coping with adversity which imposed inevitably a limit on the evolution of true brain. Even creatures of such notorious wisdom as ant and bee have had to make do with an intricate design of nerve-knots, or ganglia, in its place. Yet who would call brainless a creature which can return from a long flight and report to its community, to within a few yards, where it has discovered honey supplies? Appraised by results, the bee is a highly developed mobile unit of a sedentary brain, the hive. (Figure 1.)

Thought and personality have already been left far behind; individuality, still evident in dogs and cats, is not to be looked for below the level of the centralised brain. Individual adjustment is unknown to the totalitarian insects; it is easier to teach an octopus than an ant. Adaptation becomes a function of the community, which however also seems to be chained to the pattern of a species. Olaf Stapledon, the Virgil of science fiction, whose soaring vision once lighted on a monstrous sedentary brain, divined that its communal mechanism would not be compatible with anything that we should regard as personality.

In honour of the bee it must be admitted that the ganglion is in effect a bit of specialised brain. All creation paused for many millions of years at that stage of neural progress. A vast proportion of living things today have not reached it. The jellyfish, for instance, has neither brain nor ganglia. A nerve-net of simple cells is adequate for all its needs, the same convulsive message being repeated over and over again, simultaneously between all centres: push, push, push!

This mechanism, elementary as it is, was an important step toward the evolution of brain. It is already a responsive nerve system. Nor is amoeba, pet of the schoolroom, to be excluded from sentient and responsive life. In the presence of food nearby, amoeba helps himself just as confidently as a child reaches for a piece of cake on the teatable.

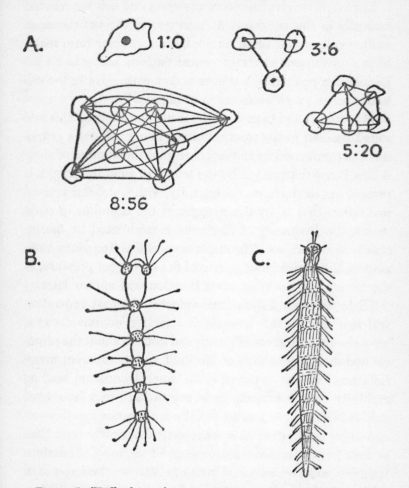

Figure 1. "Difficulties of communication increase with the number of units." Main Stages of Evolution in Organisation of Brain. (a) Nervenet (Arrested form: Jellyfish). (b) Ganglia System (Insects). (c) Notochord with Budding Brain (Lancelet).

The response mechanism of amoeba's one cell is activated basically in the same way as each one of the ten thousand million cells in our brain. Protoplasm in cellular form maintains a precarious electro-chemical balance and, like an explosive, can be fired by a trigger-action. Firing causes the cell to discharge (with an emission of impulse) and at the same time to reload and to re-cock the firing mechanism. This will sound familiar to the troops; but the living cell has a refinement not possessed by the machine gun: the number of shots it fires is not determined by the length of time the trigger is pressed but by the force, the intensity, with which it is pressed just once—that is, by the strength of the stimulus. In plain words, the frequency of discharge is modulated by the intensity of stimulation. The importance of this frequency modulation, the dependence of rate of fire on trigger pressure, in the internal economy of brain function will appear later.

Those who recall the simple amoeba of a past generation will realise from this how much scientific notions about it have changed. And consequently our notions about the physical and mechanical basis of life itself and its different manifestations. We have travelled far since Doncaster, leading authority on the subject, could say in 1919 that "the word 'cell' is beginning to lose its definite and precise significance, and to be used rather as a convenient descriptive term than as denoting a fundamental concept of biology." And since Haldane and Huxley could write in 1927 in their standard work on biology: "As a matter of fact, the only valid distinction between plants and animals is concerned with the type of foodstuffs which they can utilize."

Biology was transformed when the true properties of the cell were demonstrated. The simple cell has indeed gone; but it has gone the way of the simple atom—not quite such a long way, it is true, but far beyond the imagination of yesterday. Our ideas about them both have been transformed by a better understanding of electrical mechanics. Underlying all physiological as well as physical events we discover other events ever more complex, more subtle. Unless "programmed," we put our questions into forms which call for measured replies instead of yes or no. The answers allow us to define differences rather than proclaim distinctions.

So the "valid distinction" between plant and animal life must be a summation of differences. The main difference lies not so much in nourishment, mobility, or even basic substance, as in mechanism. The response of plant cells is mechanically different from that of animal cells: there is no transmission of impulse from cell to cell. The plant cell is a simpler form of life than the animal cell, more directly coupled to sun and earth. It may not be the simplest. Where does life begin and where is it not? The crystallographer is still exploring this field. The chemist hesitates to set up a new division between organic and inorganic matter. Many barriers have been swept away since the non-Christian biologist discarded the frontier of the soul.

But, as in that case, under a dismantled barrier are often found the traces of some real division which it may have marked or masked. We find such traces again, even while vaunting the continuity of life, as we pass with our measuring rods across the old frontier of the animal and vegetable kingdoms.

Are orchids really a lower form of life than infusoria? Are all plants immobile, all such innocent feeders? There is an appearance of reflex action in a thousand varieties. Some turn

their blossoms to the sun. Some bloom only in the dark, like
the cereus that waits on midnight or the convolvulus that un-
folds its moonlike blossoms at sunset. Venus's flytrap not only
catches its prey but devours it. Countless devices lure the
insects to pander to the flowers. There seems no reason why,
being capable of collaboration in such diverse and complex
ways, plant cells should not have evolved into conductors of
stimuli also. The very simplicity of plant life, its serenity, its
direct access to the sun's energy and earth's chemistry, might
then have been more advantageous to the evolution of brain
than the hurry-scurry of animal existence.

Moralists might like to consider whether perhaps the failure
of the plants to take advantage of these ideal conditions to
filch humanity's prize, may have been a result of sexual in-
dulgence. In many millions of years of the hothouse life which
earth provided for the young vegetable species, anything
might have happened; mutations were the order of the day
and any young plant with a spark of originality in its cells
might have struck out on a new line. What happened so
scandalously was that, with a millennial shudder of awakening not unlike our minor post-war rhapsody of a generation
ago, the plant world discovered sex. The vegetable kingdom,
hitherto blossomless, was swept by a delirium of public love-
making into the fantastic exhibitionism of the flowers; all except the cryptogams, who, as the name indicates, still cherish
the clandestinity of liaisons proper to a conservative society.

This first carnival, on the organs of which we depend for
our main sustenance, took place some 150 million years ago,
just about the time when weak little creatures began to match
their wits against the great beasts. In the struggle for survival
in a cooling world, the prodigality of sex served the plants

well. Perhaps all the better in the absence of nerves. For although it was a popular scientific notion a generation ago that plants have a nerve system, no vegetable nerve action has ever been satisfactorily demonstrated. The nearest approach to animal life is found at the lowest level, in the cells of certain algae which are so similar to nerve tissue that they have been used to corroborate experiments on nerves.

Plant cells are influenced by light, temperature, gravitation and moisture as well as contact. But plant responses to these stimuli are not neural reflex actions. When a tendril touches an object, for instance, it will curve with it and eventually encircle it, only because each cell, as it comes into contact with the object, is thereby checked in its growth, while the others untouched continue to grow, unaffected. The tendril is shaped by its environment, once for all. In the same one-way manner a plant reacts to the pull of gravity, to light and its withdrawal, to the touch of insect feet. The stimulus must be received directly by each cell. Also, the recovery of a plant cell, its recharging, is slow and uncertain. It is more like a water-pistol than a machine-gun with improved trigger action. Only in rare cases do plant cells achieve battery fire. Even then there is no modulated frequency of firing; and no plant has another shot in the locker until after a long interval for recovery, not even the most sensitive of the Mimosa family —not *M. sensitiva*, *pudica*, *casta*, *pudibunda*, or *viva*—not even *M. palpitans*.

By virtue of its unitary rhythmic explosion, the animal cell is mobile. And mobility appears to have been a main factor in the next phase of organisation.

Geneticists find it difficult to account for simple mutations. More than mutation now seems to take place in the emer-

gence of new nervous functions. Between amoeba and jellyfish is a gap. We do not know how living things passed from
the unicellular to the multicellular stage. And later on the
proliferation of nervous functions becomes bewildering. Increasing size of the brain does not explain it. Whether growth
allowed diversification or encouraged it, or was an effect of
it, or merely contemporary, we can say with assurance that
something more took place than a multiplication of cells. Not
by any stretch of imagination could our organ of imagination
have evolved simply by the accumulation, one by one, or million by million, of the necessary cells without the quite mysterious diversification of them.

Many other bodily adjustments besides increase of space
and the elaboration of its own vast communications system
had to be made for the most complex organ of all creation,
an apparatus of such proportions and such variety of function
that the giant computers with their thousands of tubes are
playthings beside it. The changes in the nervous system itself
are stupendous—differentiation of nerve tissue, diversification of cell, deviations into one specialisation after another;
lavish emergence of completely new organs or, on the other
hand, adaptation of a vestigial one in such an economical
manner that parsimony, obvious necessity in the housing of
the brain itself, appears quite unexpectedly in the architecture of other parts of the organism. And then, to cap it all,
emancipation of the brain from the very services for which
it seemed to have been millennially evolved, destined always
to serve, its sole purpose for ever.

Functionally, the only conjectural intermediary structure
between the one cell and the many is that of the sponges, if
it can be called structure. A sponge is an aggregation of uni-

cellular protozoa which have no observable cohesive agent; yet, if you tear the live sponge to shreds and even put it through a sieve, the cells will assemble again, like an organism rising from the dead. There is nothing systematic, however, about their agglomeration. If their resurrection is symbolic, their way of association is a model for the anarchist and their inter-communication a gift to telepathy.

And now a structural miracle takes place in cellular life. Part of the cell begins to stretch until it becomes specialised as an electrical conductor. It stretches until the length of the cell is as much as 100,000 times its breadth—a proportion unique in life. Once association in a true organism thus becomes possible, cells abandon extra-sensory perception and are quite materially linked in a nerve-net, as in the jellyfish. Then comes the moment—or the hidden process of many million years—in which the inherent mobility of animal cell and organism exploits the next evolutionary jump.

Anything that moves about increases its risks, runs into new dangers. The protozoan scene under the microscope is one of continual traffic jams and innumerable collisions. Jay-walkers following their insensitive noses will be eliminated in the course of a few million years. Before the days of aggressive beaks, sensitive noses have a survival value. In the multicellular world, among beings of more delicate construction than the plastic single cell, there is an even higher premium on road-sense, on steersmanship. We have to visualise an elementary system of control by which the forward part of an organism can obtain information and feed it back internally for guidance of its operative motor nerve centres. Such feedback control was a commonplace of the physiologist long before the engineer found common ground with him in "cyber-

netics," a word to which Norbert Wiener gave additional
meaning when he revived Ampère's *cybernétique*, used more
than a century ago for "science of government."

This principle of steersmanship (literal derivation of the
word) by feedback undoubtedly played a very important
evolutionary role in the first stages of animal life. Possibly
even before life appeared. Feedback as the first act of creation, or as the process of continuous creation, is a pretty subject for the metaphysician.

> Said a fisherman living at Nice:
> The way we began was like thees:
> A long way indeed back
> In Chaos rose Feedback,
> And Adam and Eve had a piece.

The young French writer on fish and philosophy, Pierre de
Latil, received this tribute to his metaphysics expounded in a
secluded Mediterranean paradise. Physiologically the theory
is conventional. Our hypothesis is that the first nervous system
evolved in this way from undifferentiated nerve-net. Some
distant cousin of jellyfish or starfish, spawned with a deformity of its nerve-net, steered its way a little more safely through
the crowded primordial sea. The abnormally congested forward units of the nerve-nets provided a dominant group of
food-and-danger predictors, overpowering with their local
majority the unorganised influence of the remaining few
nerve-filaments. So, profiting by the accident of its congenital
deformity, the worm-to-be survived a little longer than its
normal fellows and begat a race of creatures inheriting the
tangled knots of nerve in head and belly.

Up to this point the function of the nerve-net was of the
simplest. As long as its job was the transmission or non-trans-

mission of an impulse—simpler than the alternatives, yes or no—its organisation was adequate. But difficulties of communication increase with the number of units; if at the same time the variety of signals also increases, the whole organism will soon become useless and break down.

This problem of organisation is a familiar one. In 1940 thousands of us in England were concerned with a precisely similar one in the mustering of the Home Guard. Early enthusiasts will remember that our first platoons were linked in perfect equality and in such a way that any information issued by any platoon was received by all. Each platoon had its own intelligence officer. It seemed to be an excellent arrangement for the immediate purpose, a simple one. Discovery of an enemy paratrooper by any unit would at once be made known to all; also the body of guards as a whole would suffer the minimum inconvenience if a platoon were wiped out. But presently the number and variety of duties increased. If every order had to go to all, and every platoon had to send information to all the others about everything, every member of every platoon would be kept on the run, with nobody left to receive the messages or act on them. So our cheery spontaneous system, formidable as it might be for some purposes, had to give way to the traditional military centralisation of intelligence and chain of command. (See Figure 1.)

The progressive nervous system reached this point of reorganisation in the Cambrian era, some five hundred million years ago, when plant life was still emergent in its primitive multicellular form of seaweed. Fossils reveal the appearance of it in the comparatively short-lived trilobites; it can be examined in related species of that period which have remained practically unchanged to this day. In slugs and snails, limpets and starfish, the nerve-net was gathered up into centres of specialised ganglia, comparable with battalion headquarters.

It should be made clear, however, that in tracing the progress of the nervous system from one form to another, no implication of direct descent of creature to creature is intended. Jellyfish is not a descendant of amoeba, mollusc of jellyfish, or man of chimpanzee. The stems of humblest origin are indeed the farthest apart. They rise diffusely from the swarming abyss of life, in which not one but innumerable missing links have perished. Since we have learned something about the basic electrical properties of all communication mechanisms we can speculate more freely about adaptation. Functional evidence allows us to piece together what otherwise makes little sense. Only scant bits of material evidence of neural development are to be found in fossil glimpses of the past, a procession of primitive orphans and bastards of unaccountable origin. Nor shall we ever know what innumerable freaks and monsters, their mutations being expendable, fell back unseen into the abyss. The fossils leave gulfs between amoeba, jellyfish and sponge, between worm, trilobite and mollusc, between starfish, crab and insect; moreover they isolate all these from the evolutionary branch that is our main concern, greatest and most mysterious of all, the fish and other vertebrates.

Between ganglion and spinal cord, however, there is a visible bridge. The lancelets, small fishlike creatures which burrow in the sand of shallow warm waters, have a stiff neural cord the length of their body, unprotected by bony covering. For the zoologist they are a living link between invertebrate and vertebrate; and this glimpse of neural evolution has been clarified by the finding of a fossil of a similar creature that lived in the Silurian age perhaps 350 million years ago.

Millions of years before this, however, a cousin of the primeval lancelet had gone through a phase of evolution even more critical than that of the transformation of ganglia into notochord. The lamprey, living relic of the past, has a protective cartilage for its cord, its head has a rudimentary skull, and in the skull is what we must accept. in the absence of any evidence directly from the past, as the earliest model of the primeval brain.

In all there had been a labour of some thousand million years between the protozoa and the emergence of an animal with a brain. The first beneficiaries of this novel equipment, after another evolutionary jump, were fish. The special nerve services which it could provide permitted them to take advantage of fantastic mutations, many of which survive to this day. Besides fish of every size from an inch to 40 feet, there are fish of every imaginable shape, including some truly eccentric survivals, such as those which grow a beard of imitation seaweed or turn the snout into a fishing-rod.

For the electro-physiologist the most ingratiating fantasy that distinguishes the fishes from all other animals is their exploitation of the jellyfish invention, propagation of an electrical impulse. Some hundred species generate currents far in excess of domestic requirements, up to 600 volts in one species, enough to kill man or beast with its discharge. Their muscle cells are connected in series like a layer-built battery; but how this came about and what could be the use of such a power-station is a mystery. Possibly it was for defence. Some species use it for navigation in muddy waters, others for communication—pioneer telegraphists of the deep. The supply of hundreds of volts in any case is extravagant. The human brain needs to generate only one-tenth of one volt for all its complexity of services. Nor was creation lavish in that era; the ductless glands, thyroid and parathyroid, were in production then and both were fashioned out of other vestigial organs. The parsimony of nature, it will be seen later, is just as important as the wastefulness which it is so much easier to observe and to imitate.

With the fish we seem to pass for the first time beyond the frontier of utilitarian life into a realm of fancy. Any visit to an aquarium, or goggled plunge into a limpid sea, will suggest that it is a world of playful childhood. As the ways of life get more complex, play indeed becomes necessary for learning to do things upon which existence may depend. In play-pool or class-pond, the fish brought up their young to inherit a lordship of the sea which has never been disputed, even by the regressive mammals who returned to it with the additional advantages of a terrestrial education.

With brain and spinal cord securely covered, vertebrate life enters a succession of millions of years of fantastic and successful adventure, including invasion of the land; but without further important neural changes. The appearance of limbs, from this point of view, is less surprising than the disappearance of all the evidence of their evolution. Their bone was good material for fossilisation, but no fossils record the transition. For all we know, the amphibia emerged from the shallows and slime crawling on fins transformed overnight into limbs complete with their five fingers, clutching therein evidence of their starfish forbears. Late arrivals would find the land in possession of creatures more primitive neurally but more advanced in function. It was the first great test of the brain as organ of survival.

The insects were well enough equipped for a struggle. Their ancestors had failed to take advantage of whatever opportunities may have been offered by genetic freaks and deformities for advancement beyond the stage of ganglion. But they had avoided the asymmetrical heresy that led some mollusc cousins to the degradation of the gastropods. The insect ganglia were usefully distributed in pairs in their segmental bodies; but their external skeleton fatally hampered further neural development. They had no brain to match against that of the amphibia. Most of them eventually retreated, each species behind its iron curtain, into the self-sufficiency of their various totalitarian regimes.

Brain had its part in making "huge leviathans Forsake unsounded deeps to dance on sands," though the great migration must have begun with smaller creatures, abnormal offspring of fish. The Devonian and early Carboniferous beaches, one may guess, were strewn with holocausts of freaks; and not only freaks. "So careful of the type she seems, so careless of the single life." But careless enough here of types. For, as life on the land increased, the excitation of the change, the exposure to ultraviolet and cosmic rays, would produce many sports and monsters among the primeval sunbathers. There were amphibia of very great size, and every variety of shape from elephantine to serpentine. Their energy, however, seems to have been drained in adapting to so many mutations. Earth knows them today only in such survivors as the frog, a dwarf and degenerate (but educationally useful) cousin of great Mastodonsaurus, first lord of the land.

Bird and reptile, emerging, were not at first distinct but mingling in diversification, as seen in fossils of creatures that

had both bird and reptile characteristics. Their close relationship is recorded in living species; the brain of birds is an
elaboration of the grade of structure shown by that of the
crocodile. Nor were birds the Wright brothers of the animal
world; they were only practising gliding when batlike reptiles
took to the air. Not flight but feathers was their distinction—
and their salvation.

This is what was involved. The messages in a nervous system travel at speeds dependent on temperature. If the organism gets warm, the speed of signals increases, the creature
becomes excited, perhaps delirious. If it gets cold, the messages slow down, the whole organism becomes sluggish; the
creature dozes, sleeps, hibernates or dies. The birds' feathers
saved them from the fate of the improvident lords of the
earth. The penguin is so well clad that he retains his exquisite
conjugal and social awareness no matter how low the temperature.

But there is something endearing also about the early monsters. Edward Carpenter, the gentle Victorian anarchist poet,
mourns them, the web-footed beasts who had come—

(Dear beasts!)—and gone, being part of some wider plan;
Perhaps in his infinite mercy God will remove this Man!

These monsters were lords of the earth for about a hundred million years. Whether it was the heat or the eventual
cooling of their perpetual paradise that was fatal to them we
shall never know; but the sun certainly would be intoxicating,
as it always is to those who seek it from afar. Until a few million years ago their ancestors had never experienced a blood
temperature higher than that of the sea. They emerge in the
blaze of the Triassic and mature in the moist heat of the

Jurassic and Cretaceous ages. Because of their size they have been given a bad name for sluggishness and stupidity. Slow they must have been when cool nights came; on a cold morning anyone could bite off a bit of Diplodocus' tail and literally get away with it before the monster had time to look around; the signal would take so long to reach the pelvic brain, part of the creature's solution of his communications problem; still longer if it had to be referred to his other brain for choice of action. For this creature had two sets of brains,

> One in his head, the usual place,
> The other at the spinal base.
> Thus he could reason *a priori*
> As well as *a posteriori*.

The upper brain, extremely small for the size of the beast, may have been specialised for higher services, however, since so many of the lower ones were evidently taken over by its pelvic assistant. It was large enough to be aware of, and devise answers to, a small vocabulary of simple sounds used for communication—perhaps as many as those of the Hayes family ape; and for the creature to be able to learn to play. Assuredly young Diplodocus spent much time in play to get used to balancing his enormous bulk—40 tons or so—and managing his 30-foot tail.

He was a peaceful vegetarian, this greatest of the dinosaurs, supreme land giant of all time. No doubt he was careful of the huge eggs his mate produced, and took paternal interest in his young universal heirs as they disported themselves like Olympian kittens chasing their tails in slow motion. Their play must have been as funny to him, seeing them lurch and flop about in the warm shallows, as the stumble of a comedian perennially is to his audience; until their swaying steps gained confidence and rhythm and the pattern of their play acquired the dignity of a ponderous minuet.

The sex life of these monsters has been, like that of the flowers, a matter of scandal. Certainly they were subject to periods of excessive excitement, not months but millennia. They were also, in their dual brain structure, a physical prototype of schizophrenia. It has been suggested that in the hothouse Jurassic age they became racially delirious and passed into a sexual madness which led to sterility and race suicide. Another theory, that of the soldier, points to evidence of great dinosaurian armament; in defence against carnivorous enemies only those could survive who themselves became carnivorous and were protected by fantastic armour plates and spikes. Some, like the American Stegosaurus, armed heavily for defence; others, like Tyrannosaurus, fifty feet of frightfulness, for attack. But in all their lordship of more than ten thousand times the length of historical time, the dinosaurs as a race never made up their minds, never adapted racially, either to defence or attack. Looking back, in the midst of our own indecision in a similar dilemma, we can see that the question, furiously as it was contested in genetic mutations and adaptations, turned out to be completely irrelevant to their survival. They had missed something more valuable to life than the most formidable armament of either kind.

Children are still sometimes taught that the distinction between reptile and mammal is that the one is a cold-blooded animal and the other warm-blooded. This is misleading and less interesting than the facts. The blood temperature of a snake or a dinosaur beside you on a hot afternoon in the tropical sun would be many degrees higher than yours. When nights began to get cool, the activity of the dinosaurs' nervous system fell with their body temperature, only to resume in the warmth of the morning.

The emergent mammals participated in this thermal alternation of nervous activity, before their saving asset, thermostasis—the ability to maintain an even body temperature—had evolved. This nocturnal reduction of nervous energy may have been the origin of our otherwise unaccountable habit of protracted sleep. There is no evidence that it arose from any need of recuperation. But when every noon was a climax of such excitation and activity as we can observe today only in convulsions, fatigue may have been averted by the slowing down of the primeval mechanism. There will be more to say about this and other aspects of sleep later.

The acquisition of internal temperature control, thermostasis, was a supreme event in neural, indeed in all natural, history. It made possible the survival of mammals on a cooling globe. That was its general importance in evolution. Its particular importance was that it completed, in one section of the brain, an automatic system of stabilisation for the vital functions of the organism—a condition known as homeostasis. With this arrangement, other parts of the brain are left free for functions not immediately related to the vital engine or the senses, for functions surpassing the wonders of homeostasis itself.

The matter is epitomised in a famous saying of the French physiologist, Claude Bernard: *La fixité du milieu intérieur est la condition de la vie libre.* Those who had the privilege of sitting under Sir Joseph Barcroft at Cambridge owe much to him for his expansion of this dictum and its application to physiological research. We might otherwise have been scoffers; for "the free life" is not a scientific expression. He translated the saying into simple questions and guided us to the answers. "What has the organism gained," he asked, "by the constancy of temperature, constancy of hydrogen-ion concentration, constancy of water, constancy of sugar, constancy of oxygen, constancy of calcium, and the rest?" With his gift for quantitative expression, it was all in the day's work for him to demonstrate the individual intricacies of the various exquisitely balanced feedback mechanisms. But I recall in his manner a kind of modest trepidation, as if he feared we might ridicule his flight of fancy, when he gave us this illustration of homeostasis and its peculiar virtue:

> How often have I watched the ripples on the surface of a still lake made by a passing boat, noted their regularity and admired the patterns formed when two such ripple-systems meet . . *but the lake must be perfectly calm.* . . . To look for high intellectual development in a milieu whose properties have not become stabilised, is to seek . . . ripple-patterns on the surface of the stormy Atlantic.

Homeostasis is common to all mammals, with many variations in the details of its mechanisms. One of the most curious of these variations concerns *Homo sapiens.* It throws no light indeed on his origin but rather clouds the common theory. For in this respect it would seem that his ancestors were more closely related to horse than to ape. Man and horse, in offset to excessive heat, sweat; apes and other beasts do not. Only isolated and intermittent evidence of any higher significance is found in the ripple-systems of other brains than that of man. For the mammals all, homeostasis was survival; for man, emancipation. (Figure 2.)

About one prehistoric result of this emancipation we may conjecture with some assurance, man's mastery of fire. The

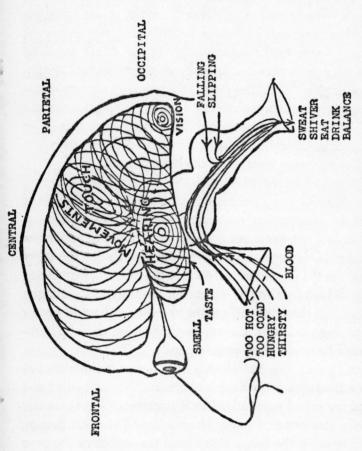

Figure 2. ". . . but the lake must be perfectly calm . . ." The upper brain is freed from the menial tasks of the body, the regulating functions being delegated to the lower brain.

37

Prometheus legend may well be the rationalisation of a tradition, just as Chinese familiarity with dragons may go right back to Pekin man's acquaintance with the last of the dinosaurs. It would be surprising if no memory, in any form, remained of this event, the dénouement of the mammalian story, the final action in the closing scene of the whole drama of man as an animal among animals, determining his final usurpation of the lordship of the earth. Whatever the circumstances of the many incidents occurring through the ages, whereby man's attitude to fire became differentiated from that of all the other animals, this must have occurred through the statistically selective function of the human brain.

Within the increased area of the cortex of the ancestral organ was evolving a mechanism capable of the new series of processes: observation, memory, comparison, evaluation, selection. As these processes evolved, man began to remember incidents of damage and danger from fire. Unlike the other animals, or the merely animal child of the proverb, man, being burnt, did not fear fire; he watched it carefully, calculatingly; he accepted the odds and sought to improve them in favour of escaping fresh hurt. He was learning which end of a burning stick he could grasp with impunity.

Man by that time may already have been a tool-maker, but even a monkey will use a stone to crack a nut; the taming and use of fire raised man indeed to Promethean heights, master not of a tool but of a force. The flinging of the first flaming brand to scare the beasts away from the mouth of the cave was a profoundly greater gesture in the human story than the dropping of the first atom bomb, its modern elaboration.

Man's mastery of fire, however, grandiose as were the horizons it illuminated, is still in the class of physical evolution;

one of the apes, with a bit of extra brain and a bit of luck, might equally well have started on the road that led to your fireside. Homeostasis allowed more than this. The perfect calm of Barcroft's lake was to be stirred by still stranger ripple-systems, whose meaningful crossing and re-crossing will be described in later chapters.

And once again, as new horizons open, we become aware of old landmarks. The experience of homeostasis, the perfect mechanical calm which it allows the brain, has been known for two or three thousand years under various appellations. It is the physiological aspect of all the perfectionist faiths— nirvana, the abstraction of the Yogi, the peace that passeth understanding, the derided "happiness that lies within"; it is a state of grace in which disorder and disease are mechanical slips and errors.

CHAPTER 2

A Mirror for the Brain

"Let's pretend there's a way of getting through into it somehow, Kitty. Let's pretend the glass has got all soft like gauze, so that we can get through. Why, it's turning into a sort of mist now, I declare! It'll be easy enough to get through——"

Through the Looking-glass

THE GREEKS had no word for it. To them the brain was merely "the thing in the head," and completely negligible. Concerned as so many of them were about man's possession of a mind, a soul, a spiritual endowment of the gods, it is strange they did not anticipate our much less enterprising philosophers of some score of centuries later, and invent at least a pocket in the head, a sensorium, to contain it. But no, the Greeks, seeking a habitation for the mind, could find no better place for it than the midriff, whose rhythmic movements seemed so closely linked with what went on in the mind.

The Hebrews also attributed special dignity to that part of the body; thence Jehovah plucked man's other self. Old ideas are not always as wide of the mark as they seem. The rhythm of breathing is closely related to mental states. The Greek word for diaphragm, *phren,* appears in such everyday words as *frenzy* and *frantic,* as well as in the discredited *phren*ology and the erudite schizo*phren*ia.

Above the midriff the classical philosophers found the vapours of the mind; below it, the humours of the feelings. Some of these ideas persisted in physiological thought until the last century and survive in the common speech of today. Hysteric refers by derivation to the womb. The four basic human temperaments were: choleric, referring to the gall bladder; phlegmatic, related to inflammation; melancholic, black bile; and sanguine, from the blood. This classification of temperaments was revived by a modern physiologist, Pavlov, to systematize his observations of learning.

As in nearly all notions that survive as long as these fossils of language have survived, there is an element of truth, of observation, in them. States of mind are certainly related to the organs and liquors designated, and may even be said in a sense to originate in them. The philosopher, William James, was responsible with Lange for a complete theory of emotion which invoked activity in the viscera as the essential precursor of deep feeling. Some of the most primitive and finest phrases in English imply this dependence of sincere or deep emotion on heart or bowels. But communication of thought is so rapid that the Greeks overlooked the existence or need of a relay station. And no doubt it is for the same reason that we all seem particularly given to the same error of over-simplification when we first begin, or refuse to begin, to consider how the mind works. We know what makes us happy or unhappy. Who, in the throes of sea-sickness, would think of dragging in the brain to account for his melancholy state?

More curious still is Greek negligence of the brain, considering their famous oracular behest, "Know thyself." Here indeed was speculation, the demand for a mirror, insistence upon a mirror. But for whom, for what? Was there, among the mysteries behind the altar, concealed perhaps in the Minerva myth, a suspicion of something more in the head than a thing, and that the organ which had to do the knowing of itself must be an organ of reflection?

The brain remained for more than two thousand years in the dark after its coming of age. When it was discovered by the anatomist, he explored it as a substance in which might be found the secret dwelling of intelligence; for by that time the mind had moved from the diaphragm to the upper story, and Shakespeare had written of the brain, "which some suppose the soul's frail dwelling-house." Dissection was high adventure in those days. Most people believed what an ironical writer today was "astonished to learn," that "it is possible for anger, envy, hatred, malice, jealousy, fear and pride, to be confined in the same highly perishable form of matter with life, intelligence, honesty, charity, patience and truth." The search for such prize packets of evil and virtue in the brain tissue, dead or alive, could only lead to disappointment. The anatomist had to be satisfied with weighing the "grey matter"—about 50 ounces for man and 5 less for woman—and making sketches of the very complicated and indeed perishable organisation of nerves and cells which his knife revealed. He could do little more. It should enlighten us at once as to the essential character of brain activity, that there was no possible understanding of the mechanism of the brain until the key to it, the electrical key, was in our hands.

There were some flashes of foresight, sparks in the scientific dark, before Galvani put his hand on the key. What generated all the speculations of the day was a new notion in science, the conception of physical motion which began to acquire importance with Galileo and continued with Newton and into our own times with Rutherford and Einstein. First among these imaginative flashes may be mentioned the novel proposal made by the 16th Century philosopher, Hobbes, when disputing the dualist theory of Descartes. The French philosopher contemplated a non-spatial mind influencing the body through the brain, and suggested the pineal gland as the rendezvous for mind and matter. The proposal advanced by Hobbes, in rejecting this popular theory, was that thought should be regarded as being produced by bodies in motion. Hobbes was born in the year of the Spanish Armada; the Royal Society had received its charter seventeen years before he died in 1679.

The controversy about the residential status of the mind is almost as much out of date as that in which the non-existence of motion seemed to be proved by the hare and tortoise fable. But the value of Hobbes' speculation was enduring; the observation and correlation of mental and physical phenomena are today a routine of physiological research.

More specific than the speculation of Hobbes was that of Dr. David Hartley about a century later. Hartley in 1749 anticipated by two hundred years the kind of theory of mental function for which evidence has been found in the last year or two. His "Observations on Man, his Frame, his Duty and his Expectations" is a milestone in the history of English thought. Hartley, a contemporary of Newton and Hume, was a pioneer of what he termed the "doctrine of mechanism." According to this, he suggested, mental phenomena are derived from rhythmic movements in the brain—vibrations, he called them; upon these is superimposed a fine structure of "vibratiuncles" which give thought and personality their subtle shades and variations. Hartley realised quite well the value of the plastic and compact virtues such a system might have. He was also the first to develop the theory of "association of ideas" in a rigorous form, relating this to his "vibratiuncles" in a manner which we should now consider strictly scientific in the sense that it is susceptible to experimental test. It is difficult for us to appreciate the originality of his notions, the gist of which is now a commonplace of electrophysiology.

Hartley wrote nearly half a century before Galvani (1737–1798) and with him we might say farewell to fancy. But to pass over the famous Galvani-Volta controversy with the bald statement that the one claimed to have discovered electricity in animals and the other its generation by metals, would be unfair to any reader who may not know how strangely truth came out of that maze of error.

The incident began with an experiment made by Luigi and Lucia Galvani in the course of their long and patient study of that still fresh mystery, electricity. The word had been in use since William Gilbert coined it in the 16th century from *elektron,* meaning amber, another pretty semantic shift; and Henry Cavendish had already, eight years before the incident, determined the identity of its dynamic laws with those of gravitation. Everybody in high society was familiar with the effects of discharges from Leyden jars upon the lifeless muscles of executed criminals; and Louis XV had, in the words of Silvanus Thomson, "caused an electric shock from a battery of Leyden jars to be administered to 700 Carthusian monks joined hand to hand, with prodigious effect." But in Bologna in 1790 the professor of anatomy had a notion that

A MIRROR FOR THE BRAIN

it was atmospheric electricity which acted upon the muscle tissues of animals. On a stormy evening, one version of the story goes, he and his wife had the curious idea of testing this point by tying a dead frog to the top of the iron balustrade of the court-yard, apparently using copper wire to hold it by the leg. They expected that, as the storm approached, the frog would be convulsed by electric shocks. And, as they watched the thunder cloud come near, so indeed it happened; the dead frog, hanging against the iron grill, twitched in repeated convulsions.

Further experiments convinced the Galvani that they had witnessed a form of electricity derived from living processes, not merely from the atmosphere. He published a famous account of his experiments on the relation of animal tissue to electricity: *De viribus Electricitatis in Motu Musculari Commentarius* (1791). Volta seized upon this to refute the whole of Galvani's thesis, repeating his experiment not only without the storm but without the frog, proving that the electricity in question could be generated by copper and zinc sheets. This "current electricity" as it was called, was therefore metallic, and no nonsense about any animal variety. So ended a controversy and a friendship. So began the science of electrical engineering.

Eppur, the Galvani might have repeated, *si muove.* For their discredited experiment had truly revealed, not indeed what they supposed, but something more wonderful. What had happened was that, swaying in the wind, the suspended frog had come into contact with the iron bars, between which and the copper wire a current had been generated, activating its muscles. The Galvani had demonstrated the electrical aspect of nervous stimulation.

This was an event as important to the physiologist as its counter-event was to the physicist; it was the starting-point of that branch of the science with which we are concerned here, electrophysiology.

Volta's counter-demonstration led directly to the invention of the electric battery, and economic opportunity evoked electrical engineering from the Voltaic pile. There was no such incentive for research when, a generation later, the existence of animal electricity was proved. Instead, the discovery was exploited by the academic dilettante and the quack. The Aristotelian doctors of the period, assuming that where there is electricity there is magnetism, saw in it proof also of Mesmer's *"Propositions"* which had been published in his *"Mémoire sur la Découverte du Magnétisme Animal"* in 1779, floundering deeper into mystification than Dr. Mesmer himself, who had at least declared in his *"Mémoire"* that he used the term analogically, and that he "made no further use of electricity or the magnet from 1776 onwards."

There is still controversy about the origin and nature of animal electricity. Nobody who has handled an electric eel will question the ability of an animal to generate a formidable voltage; and the current is demonstrably similar in effect to that of a mineral dry cell. On the other hand, there is no evidence that the electric energy in nerve cells is generated by electro-magnetic induction or by the accumulation of static charge. The bio-chemist finds a complicated substance, acetyl-choline, associated with electric changes; it would be reasonable to anticipate the presence of some such substance having a role at least as important as that of the chemicals in a Leclanché cell.

We know that living tissue has the capacity to concentrate

potassium and distinguish it from sodium, and that neural electricity results from the differential permeability of an inter-face, or cell-partition, to these elements, the inside of a cell being negatively charged, the outside positively. Whether we call this a chemical or an electrical phenomenon is rather beside the point. There would be little profit in arguing whether a flash-lamp is an electrical or chemical device; it is more electrical than an oil lamp, more chemical than a lightning flash. We shall frequently refer to changes of potential as electrical rhythms, cycles of polar changes, more explicitly electro-chemical changes. We shall be near the truth if we keep in mind that electrical changes in living tissue, the phenomena of animal electricity, are signs of chemical events, and that there is no way of distinguishing one from the other in the animal cell or in the mineral cell. The current of a nerve impulse is a sort of electro-chemical smoke-ring about two inches long travelling along the nerve at a speed of as much as 300 feet per second.

The neglect and mystification which obscured Galvani's discovery, more sterile than any controversy, forced electrophysiology into an academic backwater for some decades. A few experiments were made; for example, by Biedermann, who published a 2-volume treatise called *Electrophysiology*, and by Dubois-Reymond, who introduced Michael Faraday's induction coil into the physiological laboratory and the term faradisation as an alternative to galvanisation into the physiotherapist's vocabulary. Faraday's electrical and electrifying research began in 1831, the date also of the foundation of the British Association for the Advancement of Science; but physiology long remained a backward child of the family.

Hampered though these experimenters were by lack of trustworthy equipment—they had to construct their own galvanometers from first principles—they gradually accumulated enough facts to show that all living tissue is sensitive in some degree to electric currents and, what is perhaps more important, all living tissue generates small voltages which change dramatically when the tissue is injured or becomes active.

These experiments were not concerned with the brain; they were made on frog's legs, fish eggs, electric eels and flayed vermin. Nor could the brain be explored in that way.

> Following life through creatures you dissect,
> You lose it in the moment you detect.

It took a war to bring the opportunity of devising a technique for exploring the human brain—and two more wars to perfect it. Two medical officers of the Prussian army, wandering through the stricken field of Sedan, had the brilliant if ghoulish notion to test the effect of the Galvanic current on the exposed brains of some of the casualties. These pioneers of 1870, Fritsch and Hitzig, found that when certain areas at the side of the brain were stimulated by the current, movements took place in the opposite side of the body.

That the brain itself produces electric currents was the discovery of an English physician, R. Caton, in 1875.

This growing nucleus of knowledge was elaborated and carried further by Ferrier in experiments with the "Faradic current." Toward the end of the century there was a spate of information which suggested that the brain of animals possessed electrical properties related to those found in nerve and muscle. Prawdwicz-Neminski in 1913 produced what he called the "electro-cerebrogram" of a dog, and was the first to attempt to classify such observations.

The electrical changes in the brain, however, are minute. The experiments of all these workers were made on the exposed brains of animals. There were no means of amplification in those days, whereby the impulses reaching the exterior of the cranium could be observed or recorded, even if their presence had been suspected. On the other hand, the grosser electrical currents generated by the rhythmically contracting muscles of the heart were perceptible without amplification. Electro-cardiography became a routine clinical aid a generation before the invention of the thermionic tube made it possible to study the electrical activity of the intact human brain.

From an unexpected quarter, at the turn of the century, came an entirely new development. Turn up the section on the brain in a pre-war textbook of physiology and you will find gleanings from clinical neuro-anatomy and—Pavlov. Almost as if recapitulating the history of physiological ideas, Pavlov's work began below the midriff. He found that the process of digestion could not be understood without reference to the nervous system, and so commenced his laborious study of learning in animals.

In the gospel according to Stalin, Pavlov founded not merely a branch of physiology as Galvani had done, but a whole new science—Soviet physiology. His work indeed was original; it owed nothing to Galvani, lying quite outside electrophysiology, to which it was nevertheless eventually, though not in Pavlov's day, to contribute so much in the way of understanding.

For nearly two generations Pavlov's experiments were the major source of information on brain physiology. Workers in the English laboratories had not permitted themselves to explore further than the top of the spinal cord. One took an anatomical glance at the brain, and turned away in despair. This was not accountable to any peculiar weakness of physiological tradition but to the exigencies of scientific method itself. A discipline had been building up through the centuries which demanded that in any experiment there should be only one variable and its variations should be measurable against a controlled background. In physiology this meant that in any experiment there should be only one thing at a time under investigation—one single function, say, of an organ—and that the changes of material or function should be measurable. There seemed to be no possibility of isolating one single variable, one single mode of activity, among the myriad functions of the brain. Thus there was something like a taboo against the study of the brain. The success of Pavlov in breaking this taboo early in the century was due to his contrivance for isolating his experimental animals from all but two stimuli; his fame rests on his measurement of responses to the stimuli.

There was no easy way through the academic undergrowth of traditional electrophysiology to the electrical mechanisms underlying brain functions. The Cambridge school of electrophysiology, under a succession of dexterous and original experimenters beginning toward the end of the last century, developed its own techniques in special fields of research, particularly in the electrical signs of activity in muscles, nerves and sense organs. At the same time, the Oxford school under the leadership of Sherrington was beginning to unravel some of the problems of reflex function of the spinal cord. In both these schools the procedure adopted, to comply with the traditional requirements of scientific method, was to dissect out or isolate the organ or part of an organ to be studied. This was often carried to the extreme of isolating a single nerve fibre only a few thousandths of a millimetre in diameter, so as to eliminate all but a single functional unit.

Imagine, then, how refreshing and tantalizing were the reports from Pavlov's laboratory in Leningrad to those engaged on the meticulous dissection of invisible nerve tendrils and the analysis of the impulses which we induced them to transmit. After four years spent working literally in a cage and chained by the ankle—not for punishment but for electrical screening—enlargement came when my professor of that date, the late Sir Joseph Barcroft, assigned me to establishing a laboratory in association with a visiting pupil of Pavlov, Rosenthal. We spent a year or so on mastering the technique and improving it by the introduction of certain electronic devices. The Russian results were confirmed. To do more than this would have required staff and equipment far beyond the resources of the Cambridge laboratory.

Meanwhile, another major event in the history of physiology had taken place. Berger, in 1928, at last brought Hartley's vibrations into the laboratory and with them a method which seemed to hold out the promise of an investigation of electrical brain activity as precise as were the reflex measurements of Pavlov. When Pavlov visited England some time after we heard of this, as the English exponent of his work I had the privilege of discussing it with him on familiar terms. Among other things, I asked him if he saw any relation between the two methods of observing cerebral activity, his

method and Berger's. The latter, I was even then beginning
to suspect, might in some way provide a clue to *how* the conditioning of a reflex was effected in the brain. But Pavlov
showed no desire to look behind the scenes. He was not in
the least interested in the mechanism of cerebral events; they
just happened, and it was the happening and its consequence
that interested him, not how they happened. Soviet physiology
embalmed the body of this limited doctrine as mystically as
the body of Lenin, for the foundations of their science. The
process of conditioning reflexes has a specious affinity with
the Marxian syllogism. Others have found in the phenomena
sufficient substantiation for a gospel of Behaviourism.

Pavlov was before his time. He would have been a greater
man, his work would have been more fertile in his lifetime,
and Russian science might have been spared a labyrinthine
deviation, had the work of Berger come to acknowledgement
and fruition in his day. But again there was delay; Berger
waved the fairy wand in 1928; the transformation of Cinder-
ella was a process of years.

There were reasons for this delay. For one thing, Berger
was not a physiologist and his reports were vitiated by the
vagueness and variety of his claims and the desultory nature
of his technique. He was indeed a surprisingly unscientific
scientist, as personal acquaintance with him later confirmed.

The first occasion on which the possibilities of clinical electroencephalography were discussed in England was quite an
informal one. It was in the old Central Pathological Laboratory at the Maudsley Hospital in London, in 1929. The team
there under Professor Golla was in some difficulty about
electrical apparatus; they were trying to get some records of

the "Berger rhythm," using amplifiers with an old galvanometer that fused every time they switched on the current. Golla was anxious to use the Matthews oscillograph, then the last word in robust accuracy, to measure peripheral and central conduction times. I was still working at Cambridge under the watchful eye of Adrian and Matthews and was pleased to introduce this novelty to him and at the same time, with undergraduate superiority, put him right on a few other points. When, at lunch around the laboratory table, he referred to the recent publication of Berger's claims, I readily declared that anybody could record a wobbly line, it was a string of artefacts, even if there were anything significant in it there was nothing you could measure, and so on. Golla agreed with milder scepticism, but added: "If this new apparatus is as good as you say, it should be easy to find out whether Berger's rhythm is only artefact; and if it isn't, the frequency seems remarkably constant; surely one could measure that quite accurately." And he surmised that there would be variations of the rhythm in disease.

Cambridge still could not accept the brain as a proper study for the physiologist. The wobbly line did not convince us or anybody else at that time. Berger's "elektrenkephalograms" were almost completely disregarded. His entirely original and painstaking work received little recognition until in May, 1934, Adrian and Matthews gave the first convincing demonstration of the "Berger rhythm" to an English audience, a meeting of the Physiological Society at Cambridge.

Meanwhile, Golla was reorganising his laboratory, and his confidence in the possibilities of the Berger method was growing. When he invited me to join his research team as physiologist at the Central Pathological Laboratory, my first task was to visit the German laboratories, including particularly that of Hans Berger.

Berger, in 1935, was not regarded by his associates as in the front rank of German psychiatrists, having rather the reputation of being a crank. He seemed to me to be a modest and dignified person, full of good humour, and as unperturbed by lack of recognition as he was later by the fame it eventually brought him. But he had one fatal weakness: he was completely ignorant of the technical and physical basis of his method. He knew nothing about mechanics or electricity. This handicap made it impossible for him to correct serious shortcomings in his experiments. His method was a simple adaptation of the electrocardiographic technique by which the electrical impulses generated by the heart are recorded. At first he inserted silver wires under the subject's scalp; later he used silver foil bound to the head with a rubber bandage. Nearly always he put one electrode over the forehead and one over the back of the head; leads were taken from these to an Edelmann galvanometer, a light and sensitive "string" type of instrument, and records were taken by an assistant photographer. A potential change of one-ten-thousandth of a volt— a very modest sensitivity by present standards—could just be detected by this apparatus. Each record laboriously produced was equivalent to that of two or three seconds of modern continuous pen recording. The line did show a wobble at about 10 cycles per second. (See Figure 3.) He had lately acquired a tube amplifier to drive his galvanometer, and his pride and pleasure in the sweeping excursions of line obtained by its use were endearing.

Berger carried the matter as far as his technical handicap permitted. He had observed that the larger and more regular

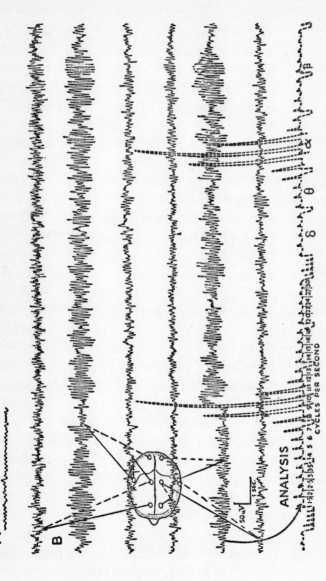

Figure 3. "The line did show a wobble at about 10 cycles per second." (a) A tracing from one of Berger's earliest records. (b) Record from a modern laboratory showing consistency of automatic analysis over 10-second periods.

55

rhythms tended to stop when the subject opened his eyes or solved some problem in mental arithmetic. This was confirmed by Adrian and Matthews with leads from electrodes on Adrian's head attached to a Matthews amplifier and ink-writing oscillograph. This superior apparatus, and a more careful location of electrodes, enabled them to go a step further and prove that the 10 cycles per second rhythm arises in the visual association areas in the occiput and not, as Berger supposed, from the whole brain.

Only some years later was it realised what an important step this was. Its significance could not be recognised while so little was known about the components of the "wobbly line," the electroencephalogram or, abbreviated, EEG. Unavoidably at the time, the significance of the salient character of the normal EEG was overlooked; it was found, in Adrian's phrase, "disappointingly constant." The attention of many early workers in electroencephalography therefore turned from normal research to the study of nervous disease. In immediate rewards this has always been a rich field. In this instance, a surprising state was soon reached wherein what might be called the electropathology of the brain was further advanced than its electrophysiology.

In the pathological laboratory, Golla's earlier surmise, that there would be variations of the rhythmic oscillation in disease, was soon verified. A technique was developed there by which the central point of the disturbance in the tissue could be accurately determined. For surgery, the immediate result of perfecting this technique was important; it made possible the location of tumours, brain injuries, or other physical damage to the brain. It was helpful in many head cases during the war as well as in daily surgical practice.

The study of epilepsy and mental disorders also began to occupy the attention of many EEG workers. The difficulties encountered in these subjects threw into prominent relief the essential complexity of the problem as compared with those of classical physiology. The hope of isolating single functions had now been abandoned; those who entered this field were committed to studying the brain as a whole organ and through it the body as a whole organism. They were therefore forced to multiply their sources of information.

It is now the general EEG practice, not only for clinical purposes, but in research, to use a number of electrodes simultaneously, indeed as many as possible and convenient. The standard make of EEG recorder has eight channels. Eight pens are simultaneously tracing lines in which the recordist, after long experience, can recognise the main components of a complex graph. The graphs can also be automatically analysed into their component frequencies. A more satisfactory method of watching the electrical changes in all the main areas, as in a moving picture, a much more informative convention than the drawing of lines, has been devised at the Burden Neurological Institute. This will be described after a simple explanation of what is meant by the rhythmic composition of the normal EEG; for its nature, rather than the methods of recording and analysing it, is of first importance for understanding what follows.

If you move a pencil amply but regularly up and down on a paper that is being drawn steadily from right to left, the result will be a regular series of curves. If at the same time the paper is moving up and down, another series of curves will be added to the line drawn. If the table is shaking, the vibration will be added to the line as a ripple. There will then be three components integrated in the one wavy line, which will begin to look something like an EEG record. The line gives a coded or conventional record of the various frequencies and amplitudes of different physical movements. In similar coded or integrated fashion the EEG line reports the frequencies and amplitudes of the electrical changes in the different parts of the brain tapped by the electrodes on the scalp, their minute currents being relayed by an amplifier to the oscillograph which activates the pens.

All EEG records contain many more components than this; some may show as many as 20 or 30 at a time in significant sizes. Actually there may be tens of thousands of impulses woven together in such a manner that only the grosser combinations are discernible.

A compound curve is of course more easily put together than taken apart. (See Figure 4.) The adequate analysis of a few inches of EEG records would require the painstaking computation of a mathematician—it might take him a week or so. The modern automatic analyser in use in most laboratories writes out the values of 24 components every 10 seconds, as well as any averaging needed over longer periods.

The electrical changes which give rise to the alternating currents of variable frequency and amplitude thus recorded arise in the cells of the brain itself; there is no question of any other power supply. The brain must be pictured as a vast aggregation of electrical cells, numerous as the stars of the Galaxy, some 10 thousand million of them, through which surge the restless tides of our electrical being relatively thousands of times more potent than the force of gravity. It is when a million or so of these cells repeatedly fire together that the rhythm of their discharge becomes measureable in frequency and amplitude.

What makes these million cells act together—or indeed what causes a single cell to discharge—is not known. We are still a long way from any explanation of these basic mechanics

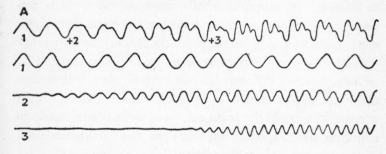

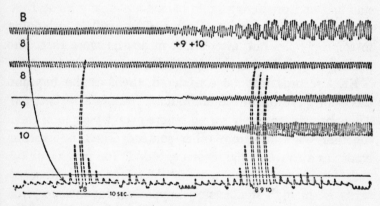

Figure 4. "A compound curve is more easily put together than taken apart." (a) A compound curve in which the three components can be detected by visual inspection, ratios 1:2 and 2:3. (b) The three components (ratios 8:9, 9:10) of this compound curve cannot be determined at sight. The bottom line shows their frequencies automatically recorded every 10 seconds. Note the accidental similarity of this curve to the EEG record of alpha rhythms in Figure 3 (b).

of the brain. Future research may well carry us, as it has carried the physicist in his attempt to understand the composition of our atomic being, into vistas of ever increasing enchantment but describable only in the convention of mathematical language. Today, as we travel from one fresh vista to another, the propriety of the language we use, the convention we adopt, becomes increasingly important. Arithmetic is an adequate language for describing the height and time of the tides, but if we want to predict their rise and fall we have to use a different language, an algebra, with its special notation and theorems. In similar fashion, the electrical waves and tides in the brain can be described adequately by counting, by arithmetic; but there are many unknown quantities when we come to the more ambitious purposes of understanding and predicting brain behaviour—many x's and y's; so it will have to have its algebra. The word is forbidding to some people; but, after all, it means no more than "the putting together of broken pieces."

EEG records may be considered, then, as the bits and pieces of a mirror for the brain, itself *speculum speculorum*. They must be carefully sorted before even trying to fit them together with bits from other sources. Their information comes as a conventional message, coded. You may crack the code, but that does not imply that the information will necessarily be of high significance. Supposing, for instance, you pick up a coded message which you think may be about a momentous political secret. In the first stage of decoding it you might ascertain that the order of frequency of the letters was *ETAONI*. This does not sound very useful information; but reference to the letter-frequency tables would assure you at least that it was a message in English and possibly intelligible.

Likewise, we watch the frequencies as well as the amplitude and origin of the brain rhythms, knowing that many earnest seekers for the truth have spent lifetimes trying to decipher what they thought were real messages, only to find that their horoscopes and alembics contained gibberish. The scientist is used to such hazards of research; it is only the ignorant and superstitious who regard him, or think he regards himself, as a magician or priest who is right about everything all the time.

Brain research has just about reached the stage where the letter frequencies of the code indicate intelligibility and their grouping significance. But there is this complication. The ordinary coded message is a sequence in time; events in the brain are not a single sequence in time—they occur in three-dimensional space, in that one bit of space which is more crowded with events than any other we can conceive. We may tap a greater number of sectors of the brain and set more pens scribbling; but the effect of this will only be to multiply the number of code signals, to the increasing embarrassment of the observer, unless the order and inter-relation of the signals can be clarified and emphasised. Redundancy is already a serious problem of the laboratory.

The function of a nervous system is to receive, correlate, store and generate many signals. A human brain is a mechanism not only far more intricate than any other but one that has a long individual history. To study such a problem in terms of frequency and amplitude as a limited function of time—in wavy lines—is at the best over-simplification. And the redundancy is indeed enormous. Information at the rate of about 3,600 amplitudes per minute may be coming through each of the eight channels during the average recording pe-

riod of 20 minutes; so the total information in a routine record
may be represented by more than half a million numbers; yet
the usual description of a record consists only of a few sen-
tences. Only rarely does an observer use more than one-
hundredth of one per cent of the available information.

"What's in a brain that ink may character . . . ?"

For combining greater clarity with greater economy, many
elaborations of methods have been adopted in clinic and
laboratory. They still do not overcome the fundamental embarrassment of redundancy and the error of over-simplification, both due to the limitations of a time scale. A promising
alternative is a machine that draws a snapshot map instead
of a long history, projecting the electrical data visually on a
spatial co-ordinate system which can be laid out so as to represent a simple map or model of the head. This moving panorama of the brain rhythms does approximate to Sherrington's
"enchanted loom where millions of flashing shuttles weave a
dissolving pattern, always a meaningful pattern though never
an abiding one." (Figure 5.)

We have called the apparatus which achieves this sort of
effect at the Burden Institute a toposcope, by reason of its
display of topographic detail. The equipment was developed
by Harold Shipton, whose imaginative engineering transformed the early models from entertainment to education.
Two of its 24 channels are for monitoring the stimuli; the
others, instead of being connected with pens, lead the electrical activity of the brain tapped by the electrodes for display
on the screens of small cathode-ray tubes. So instead of wavy
lines on a moving paper, the observer sees, to quote Sherrington again, "a sparkling field of rhythmic flashing points with
trains of travelling sparks hurrying hither and thither." As-

sembled in the display console, 22 of the tubes give a kind of Mercator's projection of the brain. Frequency, phase and time relations of the rhythms are shown in what at first appears to be a completely bewildering variety of patterns in each tube and in their ensemble. Then, as the practised eye gains familiarity with the scene, many details of brain activity are seen for the first time. A conventional pen machine is simultaneously at the disposal of the observer, synchronised so that, by turning a switch, a written record of the activity seen in any five of the tubes can be made. Another attachment is a camera with which at the same time permanent snapshot records of the display can be obtained. (Figure 6.)

Thus, from Berger's crude galvanometer to this elaborate apparatus requiring a whole room of its own, electroencephalography has progressed from a technique to a science. Its clinical benefits, by-products of free research, are acknowledged; they can be gauged by the vast multiplication of EEG laboratories. From Berger's lone clinic have sprung several hundred EEG centres—more than 50 in England alone. Literally millions of yards of paper have been covered with frantic scribblings. In every civilized country there is a special learned society devoted to the discussion of the records and to disputation on technique and theory. These societies are banded together in an International Federation, which publishes a quarterly Journal and organises international congresses.

For a science born, as it were, bastard and neglected in infancy, this is a long way to have travelled in its first quarter of a century. If it is to provide the mirror which the brain requires to see itself steadily and whole, there is still a long road ahead. The following chapters give the prospect as seen from the present milestone, assuming that such studies are allowed to continue. Looking back, we realise that the present scale of work as compared with previous physiological research is elaborate and expensive. But our annual cost of conducting planned investigations of a fundamental nature into man's supreme faculties is less than half that of one medium tank, and the money spent on brain research in all England is barely one-tenth of one per cent of the cost of the national mental health services.

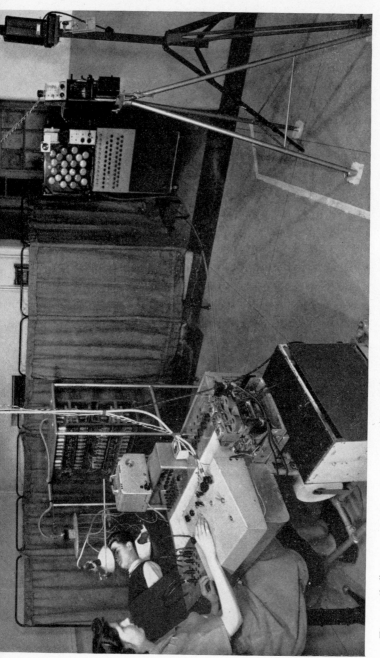

Figure 5. ". . . a moving panorama of the brain rhythms." The Toposcope Laboratory. The subject's couch and triggered stroboscope (flicker) reflector at extreme left beyond desk of 6-channel pen recorder with remote control panel. The 22-channel toposcope amplifier is in the background, the display panel at right centre, camera and projector at extreme right.

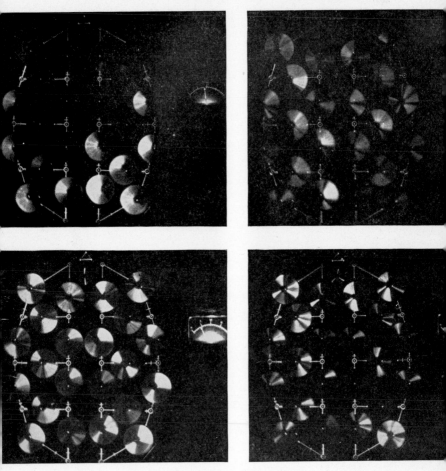

Figure 6. ". . . always a meaningful pattern though never an abiding one."
Snapshots of the "sparkling field of rhythmic flashing points." Each of the tube
screens, which form a chart of the head seen from above with nose at top,
shows by the flashing sectors of its disc the activity of the corresponding area
of the brain. (*Top left*) Resting alpha rhythms. (*Top right*) Theta rhythms in
anger. (*Bottom left*) Wide response to double flashes of light. (*Bottom right*)
Spread of response to triple flashes.

CHAPTER 3

The Significance of Pattern

Out of the earth to rest or range
Perpetual in perpetual change,
The unknown passing through the strange.

John Masefield

THE ONLY important assertion about the working of the human brain which has been discussed so far is that, in the records of its electrical activity, among their myriad patterns are found a few which seem to be meaningful. This chapter will go a step further; we shall begin to enquire into the special significance which we think may be assigned to some of these patterns. So before taking this very long step, it may be well to do two things—to clarify our outlook as strange horizons come into view, and to establish the basic axiom that where there is pattern there is significance.

The main subject throughout will be the physical counterparts of mental events. It is common experience that wherever these are discussed two attitudes toward them usually appear, contrasted, often controversial, a revival of the old dispute about mind and matter, for which each generation finds new slogans. On the one hand, some people regard these mental processes as straightforward results of complex nerve activity coupled to a complex environment; they consequently declare that human reason is easily imitated and even surpassed by electronic devices. They may go further, and suggest that it is only a question of careful observation and patience, of technical refinement, before we can observe as electrical discharges the thoughts of our own brains.

On the other hand, preconceptions aside, and considering only the physical intricacy of the brain, its vast aggregate of cell units, the still vaster astronomical number required to express their combinations and permutations, the boundless variety of human thought and imagination—considering only these factors of consciousness and personality, one may well question how, or even whether, such matters can ever be included in a philosophy of reason. They may seem to belong to another "universe of discourse," wholly inaccessible to our technical probing, not merely because, like the other side of the moon, they happen to be facing the other way, but because, like a solid to an inhabitant of Flatland, they exist in a dimension which cannot by its nature be indicated on our measuring instruments.

Psychiatrists recognise these opposites in their patients and political theorists in historical events. One should appreciate also that the conflict is not only between individuals, but in one individual from time to time; as well as controversy there is vacillation. A general reason why this is so—why the brain-mind relation is a natural subject of controversy and vacillation—may be noted. Mental functions are active adaptations, but past studies of the brain have nearly always been static. The psychologist, rushing in, asked how the mind works while the physiologist too modestly was still asking what the brain is. The situation has changed in these ten years. But the physiologist today, dealing with an adapting mechanism, knows that he will not get exactly the same answer twice to his continuous questioning and that no demonstrable truth is both absolute and general.

So he must reject the alternatives of the controversy. For him neither a glib materialistic fantasy of the allknowing, nor defeatism of reason in evasion of the difficulties presented by the complexity of the subject. There will be plenty of reason, on the one hand, for being modest about what has been discovered; every discovery leads to new horizons of the unknown. And as for the complexity, that is only the background of life itself. Foremost here will be the discovery of simplicity and parsimony even in the richest of life's manifestations.

Consider then the significance of pattern. So much of brain physiology rests on this conception that it must be thoroughly understood before going further. It would be grossly misleading, for instance, to let it be thought, as loose talk about "brain waves" has suggested, that there is any reason to suppose *a priori* that the recorded patterns of brain activity have any mental significance. Indeed the reverse is the case; the alpha rhythms, most prominent of all those patterns, are most regular in their formation when the mental activity is least.

"Order is Heav'n's first law," wrote Pope. Pattern is the raw material of order. It is not always so regarded, nor even universally recognised as significant. We have a way of importing words for one purpose and then using them for another and quite different one, to the glory of English and the confusion of foreigners. Pattern is one of them. The word came from Italy by way of France—*pater, patron;* but pattern is something which cannot be named in Rome or Paris, for which there is no word in the Romance languages. Their dictionaries can only guess at it with words strictly meaning model, design, sample, exemplar, and so forth, as misleading as precision could be. Perhaps they had no need of such a word, so imbued was the Latin tradition with pattern; the Romans could make regular divisions of their daylight regardless of its duration, as if the seasonal variations were, like the tides of their sea, negligible, and the man-made pattern of a day the whole matter. For us, creatures of the glacial periphery, there was no escaping the majestic variations of light and darkness, incalculable but meaningful rhythm. The extravagances of its millennial flicker are imprinted in our nature; witness the brilliant aberrations of Baltic mentality where supermen loom in the twilight of the gods.

Regular or irregular, the alternation of day and night was probably the first pattern in time that impressed the human brain, accentuated by the pre-thermostasis imposition of sleep. There would be early recognition, too, of the patterned calls of animals, the warning sequence of a coming storm, memorable cries of defiance or distress; finally, most precious of all patterns in time, speech. Long before this, however, observation and memory of patterns in space would be highly developed, though limited in practice to two dimensions— the relative position of places, spoors, waterholes, caves. Early in our history we must have found safety and satisfaction in remembering and reproducing simple patterns.

Pattern, then, may be defined as any sequence of events in time, or any set of objects in space, distinguishable from or comparable with another sequence or set. The first significant attribute of a pattern is that you can remember it and compare it with another pattern. This is what distinguishes it from random events or chaos. For the notion of random—another imported word lost to the Romance languages—implies that disorder is beyond comparison; you cannot remember chaos or compare one chaos with another; there is no plural of the word. Pattern is a quality of familiar things and familiar things are comforting things. It is much nearer the truth to say that man abhors chaos than that nature abhors a vacuum. Man seeks patterns for his comfort and, in English at least, can boast himself a pattern-making animal.

Broadly speaking one may say that the sciences derive from pattern-seeking, the arts from pattern-making, though there is a much more intimate relation between the seeking and making of patterns than this would suggest. It is worth recalling some other instances of pattern-seeking as the origin of a science.

When man first looked at the night sky from the nursery window of the race, he found patterns in the disorderly array of the stars, and his picturesque constellations are still useful to the astronomer, the beginning of his science. Geometry, the first abstract science, arose from the need for the seasonal restoration in the Nile Valley of the patterns of property. Gravitational astronomy, as well as much else, grew from Newton's perception of a pattern in the mutual relations of physical bodies. Botany rests on the pattern of plant relations which Linnaeus transformed into a classification. Biologists are still trying to fill in and complete the pattern of species Darwin found in the animal world. In later times, Mendel discovered the pattern of heredity, the science of genetics. Clerk Maxwell, unrecognised a hundred years ago except as a competent mathematician, endowed this generation with his prophetic enunciation of the patterns of electrical forces without knowledge of which there would be no electronics.

Knowing well this eagerness to discern and the temptation to imagine pattern, the French father of physiology, Claude Bernard, deprived of this particular word, resisted and decried adherence to any "system." A scientist, he warned, should realise that "systems do not exist in nature but only in men's minds." The paradox that "philosophy should not be systematic" was snatched by Bernard's compatriot Bergson sixty years later as a cornerstone of his philosophy; the irony of such an inveterate pattern-seeker as Bernard concluding that either men's minds were outside nature or nature is chaotic could not have been more nicely pointed.

Finally, among these Fathers of Science, Berger has a curious position as a candid seeker for pattern in the electrical activity of his brain, audaciously neglecting Bernard's assertions that he should seek it only in his mind.

Without pattern-making also, however, none of these brilliant perceptions of pattern would have been possible, let alone communicable. In every case there were only fractional pieces of information available, with many other bits of unknown quantities, as in the illustration of the cryptographic nature of EEG records in the preceding chapter. The putting together of these broken bits of information and questions, of known and unknown quantities, was made possible in the first instance by the Arab invention of patterns in which visible and invisible pieces could be logically manipulated, assembled, equated, evaluated—in a word, algebra.

Classical physiology, as we have seen, tolerated only one unknown quantity in its equations—in any experiment there could be only one thing at a time under investigation. This was arithmetical physiology: a single number was extracted from a known sequence, examined, altered, replaced, and the effect on the sequence noted. The EEG pattern is never simple or singular. We cannot extract one independent variable in the classical manner; we have to deal with the interaction of many unknowns and variables, all the time. We have to deal with the rhythms algebraically.

In practice, this implies that not one but many—as many as possible—observations must be made at once and compared with one another, and that whenever possible a simple known variable should be used to modify the several complex unknowns so that their tendencies and interdependence can be assessed. This is an operation analogous to the solution of simultaneous equations; its practical side will be described in detail when discussing the purposes and effects of flicker stimulation. We must first consider the main varieties of spontaneous pattern observed and how they reach the brain from the different sense organs.

The anatomy and minute structure of the nervous system is so intricate, and it is so well described elsewhere, that the reader who must envisage a structure in order to imagine a function would do well to consult one of the works listed in the bibliography. Physiologists find convenient a division of the whole nervous system into two main parts, the sensory, receptor or afferent mechanisms, and the motor, effector or efferent ones. On the sensory side are included the sense organs, the sensory nerve trunks, each containing myriads of minute nerve fibres, the extension relays of these within the spinal cord and, terminating at the surface of the back half of the brain, the sensory cortex. On the other side of the divide is the motor cortex, in the front part of the brain, the motor nerves and the muscles and glands which act upon the body itself or upon the outside world. There is here a complete circuit of information and effect: event, receptor, nerve, brain, brain, nerve, effector, event. So dramatic are the properties of such a circuit that for many years physiologists were content to study its simplest functions. Anatomists, too, sought tirelessly for the structural characters which were expected to underlie and to mirror the simple functional pattern outlined by experiment. But, just as the discredited phrenologists —who filched a good name from the science we are introducing—found too much meaning in their bumps, so some microscopists and, it must be confessed, electrophysiologists, saw system in the brain where others found none, and mistook the constriction of their methods for consistency in their material.

It is true that certain parts of the brain have a regular and recognisable microscopic appearance and respond in a fairly predictable fashion when stimulated electrically, and when diseased or damaged they are associated with certain symptomatic or diagnostic signs or symptoms. But the exceptions to these rules are so numerous and their experimental foundation is so tenuous, that there is now a tendency to support an entirely "holistic" view of brain function, to suppose that all parts are engaged in any sense or any action, and that the location of function is more a probability than a place. In such a system, pattern there still must be, but not a pattern that can be recognised at a glance or described in a phrase. That we can conceive of it at all is due to the obscure workings of the brain regions which yield least to experimental probing, the association areas, sometimes called "silent" because their oracles are dumb when threatened by the experimental intruder. These regions make up the greater part of the human brain and are closely linked with the receiving areas where impulses from the receptors most commonly arrive.

The sensory reception areas or departments, we have suggested, are more like leaseholds than freeholds of the senses, so vagrant are the functions of the brain. These sensory systems, before anyone knew how they worked, used to be compared with telephone exchanges, simply because messages were known to travel between the end organs and the brain. Taste, least important of the senses, could then be conveniently described as a local exchange with very few outside wires and only four numbers of its own—sweet, sour, bitter and salt. But the analogy is just about as misleading as it could be. This must be explained.

In a telephone system the meaning of a message received depends on the sender; in a sensory system the meaning depends on the receiver. When nerve impulses travel from a sense organ it is their destination on the cortex which determines, in the first place, the character of the sensation, not the sense organ from which they come. If, when you get a number on the telephone, you give a message, the message remains the same, even if you give it to a wrong number. The result of such an error in the brain is very different. Supposing some vinegar comes in contact with one of the sensitive end organs of taste in the tip of your tongue and "gets a wrong number"—that is, say, supposing the nerve fibre conducting the impulse provoked by the vinegar, instead of connecting with its proper reception area, becomes in some way cut and grafted onto a nerve fibre leading from the ear to the brain— what do you think you would taste? You would taste nothing. You would hear a very loud and startling noise. Every time the nerve end in your tongue was stimulated you would have a similar hallucination. If, instead, one auditory nerve were in this way misconnected with an optic nerve, when you heard music you would see visions. This is the mechanistic basis of hallucination. Such accidents are unlikely to occur in the peripheral nerves, where cross-talk is avoided by special anatomical arrangements, but they do occur within the brain itself.

The matter is not, of course, quite as simple as here outlined; nor will the reader look for a complete account of the mechanisms of the senses in what follows. But it seems to be a safe generalisation, about all the senses, to say that what we usually speak of as the "quality" of a sensation depends upon the parts of the brain reached by the nerve impulses involved, and that the intensity of this sensation depends on the frequency of the impulses.

Smell, the better part of taste, is more interesting because it has a greater variety of sensations and many more associative links with the activity patterns of the brain and the emotions. The reactions of the olfactory organ to molecules which have a distinctive smell can be observed as impulses travelling from it to the brain. Their pattern in an EEG record shows that frequency of impulse increases with intensity of excitation. Just how the brain discriminates between different smells is not so clear. Adrian, who led exploration in this field, found slight but distinct difference in the sensitivity of different receptors to different kinds of smell, and indications that the impulses from the olfactory organ are relayed from specialised receiving areas to the main sensory receiving area for the nose and mouth on the lateral surface of the cerebral cortex. Thus it is not yet clear whether the whole pattern of smell is relayed to one part of the cortex or if the different components of it go to different parts and are reassembled with discrimination at a later stage.

Touch takes up a considerable area in what may be described as a much folded and wavering chart in the cortex. This is to be expected, considering the extent of discrimination which is required for us to know where a tactile message originates. All the organs and parts of the body are represented on the chart, not in their visible proportions but according to the neural equipment and importance of each part —mouth larger than the cheeks, fingers larger than the arms, and so forth. The web of patterns woven by the many end organs on this shimmering abstract of our flesh is so intricately entangled that it may be some time before any satisfactory study of them will be made. Nor is the complexity quantitative only.

The sense of touch is the most vital of the senses if only because of its major role in reproduction. And here we come upon a limitation of the brain's emancipation, in which some physiological support for Freud's thesis may be found regarding the intrusive mental character of sex. The operational isolation of sexual function is peculiar. Intercourse and procreation are not necessarily conscious acts; nor are they naturally effected on the direct initiative of the higher centres; yet there is constant access for sensory stimuli from the organs to the brain. In this the brain is rather like an intelligence unit in the field which can receive information but cannot transmit any tactical orders. Strategy, of course, is another matter, and so are discipline and art. Many civilised peoples have sought to endow sex with a ritual. This has at least an effect of providing a pattern of sex behaviour for the brain.

Physiologically considered, some special art or discipline of this kind would be preferable to the mild but unceasing stimulus of the popular arts of today, excitatory without being formative. Heaven only knows what mutations this titillation may in time promote, neither so lordly as those of the amphibia nor so decorative as those of the plants. Puritan and sensualist agree that the pleasures of touch, unguided by any art of their own, are nasty, brutish and short.

In this sector something might be learnt from the blind about the intelligent uses, artistic as well as utilitarian, of the sense of touch. The person who obtains an image of a spatial object by touching it, or scans the plastic symbol of a verbal code, adds to the transformations of temporal and spatial pattern, and does so with a refinement and concentration which cannot fail to be significant. In most people the sense of touch and feeling is rudimentary, ambiguous, inconsistent and subordinate.

Pain is often erroneously regarded as an unpleasant exaggeration of the sense of touch. In fact it is more like a sixth sense. It has its own interior and exterior end organs and its exclusive nerve connections. Messages pass along them at a much slower pace than those on the touch and other sensory systems. The White Queen was wiser than Alice thought when she screamed before she was hurt instead of after. This is what, in effect, we always do. If you step on a tack, you jump first, and only later feel the pain of it. The reflex signals travel there and back in less time than it takes for the sensation of pain to reach the brain. Try it. You'll be surprised.

Patterns of hearing are second in interest only to those of the eye. Any sound—be it a single note or many simultaneous notes that make mere noise or combine in a musical chord—

is an event in time. The spatial aspect of a chord is only symbolical. A note must have a minimum duration; it must last long enough for its wave-pattern to repeat, like that of wallpaper, for identification. The lower the note, the longer the duration must be; a chord too briefly sounded loses its bass. And the pattern of sound with which ear and brain have to deal has other qualities besides duration or rhythm; pitch, timbre and amplitude must also be identified. But before a pattern of sound reaching the ear can be transmitted for spatial distribution on the cortex, its components must be identified. The chord must be analysed.

This is done by the cochlea, a small coiled instrument of the inner ear, not to be confused with the comparatively unimportant ear-drum. The coiled tubes in the cochlea are less than an inch long, but they are so interconnected that the fluid in different parts of them vibrates in response to every pitch of audible tone or harmonic. Vibrations of a given frequency activate one section of the tube to which a responsive membrane is attached, and from this section of the tube a nerve fibre transmits a signal to a point on the reception area of the cortex. Thus the pitch, the wave frequency of a note, is converted into a single pinpoint for spatial discrimination, as are also any harmonics of its timbre. Its characterisation by wave-frequency ceases at the cochlea. What is indicated by the frequency of the nerve impulses is the amplitude or volume of the note; the louder the sound, the more rapidly the impulse is repeated.

There will be more to say later about the sensitivity of the sleeping brain to sound, as compared with light or any other sensation. Music is proverbially connected with the emotions and provides a thousand examples, among the daily patterns of our lives, that familiar things are comforting things. Some rhythms invigorate the normal activity of the brain; others may suspend or dissipate it. The objection to constant broadcast streams of light music is not that the music is light but that it is hypnotic and its associations maintain a constant pattern of mild titillation.

The reader can test the brain's suggestibility and its affinity to rhythm. Listen to the regular ticking of a clock, and note how long it takes for the unaccented sequence to resolve itself into groups of two, three, or four ticks. This affinity to rhythm, the close correlation between external patterns of sound and those within the brain, suggests one reason why loss of hearing, for which no other sense can deputise, is so hard to bear; another reason may well be that the ears have no natural experience of deafness, nor any mechanism for temporarily excluding the audible world, as darkness or closing the eyes makes us familiar with a transient blindness.

The pre-eminence of vision is supported by organs of supreme elaboration and sensitivity, and by the impressive proportion of the activity of the brain which is devoted to that sense. It is here that the brain's emancipation from the menial services of the senses becomes most apparent. All the intricate subsidiary operations of vision have been relegated to the lower centres. The neural situation may be compared with that of a film studio. The director is not concerned with the details of operation of the cameras; focus, aperture, exposure —these are expertly attended to without any orders from him; but their positioning and aiming are a great part of his executive skill, and the choice of set or location and working hours is under his authority. He himself is under no compulsion to note what is being taken. How often in our own experience he simply does not see what we are looking at! Seen by him or not, the film is shown in the projection chamber; the picture reaches the projection area of the cortex. Seen or unseen, it goes to the archives; the whole thing, from beginning to end, may be available in the mysterious cabinet of the memory, long after we have, unreminded, forgotten all about it.

The analogy of photography breaks down, however, when we come to look at the mechanism of vision. To illustrate the difference, let us consider what happens when we look at the most simple scene, a scene in which all the details remain the same for a long time, a true space pattern, say a landscape. You may think that you can take in the whole of a static scene of this kind at a single glance, as a camera does, without moving your eyes. But that is not how the eyes work. An instantaneous glance will not allow your retina to show you an image like that obtained by a camera. The most your retina will obtain from a swift single glimpse will be a very small clear centre in the midst of a large field of quite indistinguishable details.

The explanation of this is simple. It is true that the lens of the eye is like that of a cheap camera, but the retina on which the image is projected is quite unlike a photographic plate or film in one important respect. The whole area of a photographic film has a uniform grain; the chemically sensitive specks of matter composing it are equi-distant and equi-sensitive. The surface of the retina is not uniform in this way. Only a minute patch, about one-third of a millimetre wide, in the centre of the retina, has a grain fine enough to receive and transmit an image in great detail. This patch contains special light-sensitive cells or cones with separate individual nerve fibres leading to the brain. Around this patch are groups of other cones, and the more sensitive but sparser and less discriminating rods, which have a common nerve fibre only for each bunch as their link with the brain. This arrangement therefore provides a high capacity for perception of detail in the very limited centre of the scene to which the eye is directed, with very poor discrimination but greater sensitivity elsewhere.

The reader can test the difference between the central and surrounding acuity of the eye without putting down this book. Turn to a new page, and as you do so fix your gaze immediately and steadily on the first word you pick out in the centre of the page. If the eyes are not moved at all it will be found that only one other word or so can be read; all the rest of the page, though visible, will be indistinct and quite unreadable. It is not just a matter of focus. To read this page of print you are scanning it, line by line, your eyes are moving so as to bring the words in sequence into the centre of the field; images of them are then received on the one minute patch of the retina which has fine enough grain to resolve the detail for recognition.

The proportion of the visual field which can receive a precision image is only about one-hundredth of the whole. The angle subtended by the minute patch is about two degrees out of 180, the visible field being something more than a semicircle. So in order to view that landscape which you may have thought you could take in at a single glance, you have to make several hundred peeps and sweeps, requiring thousands of co-ordinated eye movements to scan the scene. With the additional effort required for discrimination of colour and third-dimensional position, this sounds tiring. And physiologically it is. For the nervous system, looking at a picture of a land-scape is vastly easier than looking at the landscape itself. The reproduction of a scene that in nature extends across half our field of vision, reduced in size to subtend a smaller angle, say, ten degrees, and flattened and shorn of irrevelant details, saves an enormous amount of effort. The area to be scanned is reduced to about one hundredth of its natural size. No doubt the lessening of the strain has something to do with the particular pleasure we feel in looking at a painting or a photograph.

The next stage in the process of vision is the transfer of the image on the retina to the projection area of the brain. This is a section of the cortex not more than about two inches square, tucked away in the back of the head. It is connected with the retina by the nerve fibres of the rods and cones, about one million of them comprising the optic nerve. The whole image on the retina is transmitted at the same time to the cortex. What the scanning of the eye does is to bring the different parts of the scene in turn onto the macula, the minute discriminating patch of the retina.

Transmission of the image from retina to cortex is continuous, but the retina is not an entirely faithful transmitter. Every image it receives persists for a tenth of a second. So if there is a rapid succession of changing images, the after-effect of one overlaps the succeeding one. It is this well-known phenomenon of the persistence of vision that makes it difficult to see the detail of rapid movement, but also makes a rapidly flickering light seem continuous.

It may be recalled that this overlapping effect is oddly connected with horse-racing, and that the connection had a result of the most widespread and popular interest. For a long time it was disputed whether a galloping horse ever has all its feet off the ground at one time. The eye could not follow the movements continuously, owing to persistence of vision; nor could it stop to analyse them. To settle the issue, a succession of photographs was taken with very short intervals. This showed that the horse was in fact airborne for a fraction of a second; but, more important to this generation, it was found that when the photographs were looked at in rapid succession there was an illusion of a moving picture.

The process of vision is not complete, however, when the image reaches projection. Every sensation brought to the cortex by the million nerve fibres from the retina must be made available for cognition to other parts of the brain. Obviously this is a greater mechanical problem than the discrimination of the limited number of signals of the other senses. A moment's consideration will also show that the provision of direct neural communication for the million visual units of the cortex with the rest of the ten thousand million units of the brain would strain the housing capacity of the cranium. The number of fibres would have to be of the order of $10^6 \times 10^{10}$ or, to write it out for the unmathematical, 10,000,000,000,000,000 fibres. A house, let alone a human head, would not contain them.

Chance provided a hint about a possible solution of this problem, and the clue was followed in the laboratory with some satisfaction. The interest of the trail invites resumption of a narrative which was broken off at a point in the story of the discovery of the brain where the next chapter may conveniently begin.

CHAPTER 4

Revelation by Flicker

Between the conception
And the creation
Between the emotion
And the response
Falls the Shadow.

T. S. Eliot

EVOLUTION FROM cell to cerebrum required some 2,000 million years; the recognition of brain by brain came only after several thousand years of intellectual maturity; the acceptance of it as an organ proper to physiological study, in spite of its formidable complexity, is recent. The next stage of its progressive revelations passes quickly through the narrow road in which research inevitably found itself in war years, into the open spaces of exploration, with a new experimental approach to the peaks of thought and personality already showing above the horizon.

The 1939 war was on us before we had time to make much of a systematic survey of the vast field which Pavlov and Berger had opened to our young physiological eyes. Four phases of development in the study of the brain since their days can be distinguished.

1. The first period, from 1928 to 1935, was one of general scepticism, neglect of Berger's original observations, undervaluation of Pavlov's exploration, disregard of any possible

connection between them. The technical devices of the period were so poor, and Berger's data were so poorly presented, that this cannot be regarded as surprising.

2. The second phase began when Adrian and Matthews, by more careful experiment with better equipment, confirmed Berger's claims in general about "the alpha rhythm." This put the seal of academic approval on the study of the brain. At the same time, Golla's surmise, expressed in 1929, that changes in the electrical rhythms of the brain would have diagnostic value, was confirmed by a study of clinical and experimental material at the Central Pathological Laboratory and the Maida Vale Hospital. This brought identification of a different sort of electrical activity of immediate clinical and later research importance. A good name for this discovery seemed to be delta rhythm because of its association with disease, degeneration and death. And with defence; for the association of these slower rhythms with disease does not seem to be accidental, but rather a sign of the mobilisation of organised defence. This is the activity by which, we found, the site of a brain tumour can be located without inconvenience to the patient. It arises in regions which are compressed or invaded by growths or distorted by injury; but it is the normal and indeed dominant rhythm in the first year of life and in sleep. (See Figure 7.)

This clinical success for EEG, with other discoveries holding promise of breaking through some of the mysteries of epilepsy, was great enough to distract nearly all workers from the fundamental problems toward the close of this period. And this, in one way, was fortunate—as a preparation for the technological interregnum of war. Two other novelties of this period, electric shock therapy and leucotomy, indicate also the radical change of attitude to the brain that was taking place. Their main interest to us here is their physical interference with personality, and they will be discussed later in that connection.

3. The war period called for a rapid development of brain technique as a clinical accessory and for the design and manu-

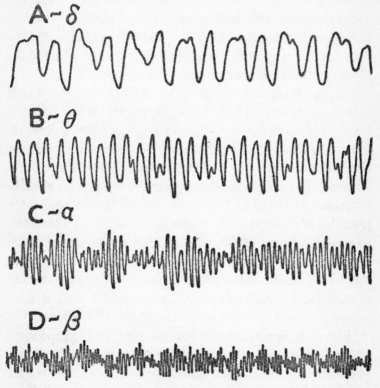

Figure 7. ". . . the frequency of a rhythm is more significant than its amplitude . . ." Main Types of Brain Rhythms. Records showing the principal wave-forms found in EEGs. (a) Delta—0.5 to 3.5 cycles per second. (b) Theta—4 to 7 c/s. (c) Alpha—8 to 13 c/s. (d) Higher Frequency (Beta)—14 to 30 c/s.

facture of better EEG equipment, the training of assistants and the formation of special societies—developments which probably would not have occurred until much later if physiological research had been uppermost. There were two EEG laboratories in the United Kingdom at the beginning, fifty at the end of the war. The first National EEG Society was formed primarily to make available our fund of knowledge to centres dealing with military casualties. At the first meeting of this group, held literally in the midst of the battle of London, criteria of electrical abnormality were agreed on and a glossary of terms was drawn up.

But the taking and interpretation of EEG records is a delicate and complicated task; it calls for something like the combined training of radiographer, who studies for two years before taking X-rays, and radiologist, who takes a long medical course before assuming diagnostic responsibility. There is not, it is true, the same physical danger in the one process as in the other; the taking of an EEG is innocuous; but a wrong interpretation of a record can be worse than burning. It was mainly as a necessary aid in this situation that we developed at the Burden Neurological Institute a device capable of decoding the EEG records automatically, plainly displaying the frequencies and amplitudes every ten seconds, a kind of wave analysis, corresponding roughly to a Fourier Transform.

With this apparatus the various electrical components, present at the same time in any given area of the brain, are automatically filtered out and displayed separately on the record, rather as the colours in white light are separated and projected by a prism. This greatly facilitates interpretation, besides providing research with a powerful instrument. For instance, one direct result of its use in 1943 was that we were able to identify yet another rhythmic component, which we named the theta rhythm; about this there will be much to say when we approach the distant Everest peak of personality.

4. The fourth and present stage of development began when the end of the war brought release for fundamental research. It is with the explorations and ventures of this period that the rest of this book will be dealing.

The equipment used today for studying brain activity contains many electronic parts and devices which were developed for radar apparatus during the war. An EEG recorder usually has over a hundred tubes, resistances, condensers and so forth, with many rows of calibrating and operating knobs and switches. Its formidable and intricate appearance not infrequently prompts the uninitiated to ask whether such a display of ingenuity is really necessary. But if we consider the complexity of the object which it is designed and constructed to examine, the most elaborate EEG equipment can only be regarded as comparatively simple in design—and extremely coarse and clumsy in construction.

The basic principle of it, at least, is simple enough. Electrodes—meaning no more than the derivation of the word, ways for electricity to pass—are firmly held to the scalp, where they pick up the minute voltages due to the electrical fluctuation in the brain. These may be from all parts of the brain, but the rhythms from the regions nearest to a given pair of electrodes are most conspicuous in the amplifier channel connected to that pair of electrodes.

Significant discharges can be recognised which are as small as 5 microvolts—that is, 5 millionths of a volt; the average is higher than this, but anything above 50 microvolts is unusual except in children. With the automatic analyser the recognition of still smaller voltages is possible. (See Figure 8.) Sustained rhythmic activity at an amplitude of only one-tenth of a microvolt can be clearly distinguished against the background of random noise. For some notion of the minuteness of such a faint current it may be reckoned how many of them would have to combine their voltage to light the common hand torch; 30 million of them might do it. The usual order of amplification for recording and other observation is four million times.

The frequency of rhythm, the number of times a cycle of its pattern is repeated in a second, is more significant than its amplitude or voltage. The delta rhythm mentioned above has a frequency of 0.5 to 3.5 cycles per second; the theta rhythm, 4 to 7 c/s; and the alpha rhythms, most prominent of all in area and amplitude in normal adults, have a frequency of 8 to 13 c/s. (See Figure 7.)

Any discovery is liable to be neglected or to give rise to strange theories to account for new phenomena, whichever branch of science may be concerned. Discoveries in physiology are particularly vulnerable. This is not because physiologists are more sceptical or more gullible than other scientists, but because their science is different. Until recently, whenever the functioning of any nerve or organ had been demonstrated in a few experiments, it was assumed that it would be the same in any other individual. So, after years of neglect, Berger's discovery became the subject of theories based on inadequate verification. By classical standards the two or three subjects examined should have been sufficient for pronouncement of a theory; today we know that there is so much variation between individuals that we must have two

or three thousand samples before any new bit of information
about the brain can be usefully incorporated in a hypothesis,
and that many more thousands must be examined and compared before venturing on a general theory. The clerical work
is therefore a problem in itself. The adoption of statistical
routine clinical work requires the introduction of an elaborate

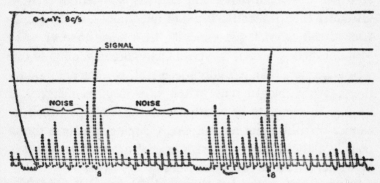

Figure 8. ". . . the recognition of still smaller voltages is possible." The top line is a record of a signal of one ten-millionth of a volt,
indistinguishable against its background of "dismal universal hiss," but
detected by the analyser and its frequency displayed far above the
noise level.

punched card-index system in which every card can hold
any number of two hundred classified items of information
about the case, and all the cards containing any given one or
any combination of those items can be assembled automatically. Thus the laboratory's functions of memory and association begin to approximate crudely to those of the brain itself.
Research was resumed, then, after the war, with a great
body of knowledge on hand which had come into existence
not quite so crudely as that obtained on the battlefield of an
earlier war, but incidental to military events rather than as

the result of planned scientific endeavour. There had to be much sifting of information, much refinement of instrumental technique. And a corresponding improvement of statistical method; for, like any other system of magnification, that of the thermionic amplification of minute voltages has its limitations. Just as the resolution of a microscope ends at the point at which Brownian movement and the wavelength of light introduce their respective types of uncertainty, so the amplifiers of the physiologist generate their undertone of what Milton called "a dismal universal hiss." Circumvention of this cannot be more than partially successful. At some point therefore the physiologist must patch with inspired imagery, if he is a Sherrington, or with guesswork, the gaps which he cannot span by exact measurement, hoping of course that a bold hypothesis may suggest a crucial experiment that will be within the range of his resources.

It was rather in this happy speculative mood that the possibility of finding out more about brain rhythms, by imposing other patterns on the brain through the senses, was suggested. Very few of the factors affecting the spontaneous rhythms were under the observation or control of experimenter or subject. Usually only the effects of opening and closing the eyes, of doing mental arithmetic, of overbreathing and of changes in the blood sugar were recorded. Sometimes the influence of drowsiness and drugs could be investigated. But the range and variety of methods were not comparable with the scope and sensitivity of the organ studied, and the information obtained by them was patchy in the extreme. The only unequivocal answers obtained by EEG had been to clinical problems; the more subtle and transient variations in the normal state of the brain escaped notice or had to be dismissed because their importance could have been estimated only if a detailed continuous assessment of the general state of the subject had been available.

In 1946 we found that the information contained in EEG records could be greatly increased by subjecting the brain to rhythmic stimulation, particularly by the flickering of a powerlight in the eyes, open or closed. Early experiments of this sort had been done by shining a light through a rotating wheel with wide spokes, but the results had been inconsistent. One trouble was that as the frequency of the flicker was increased by turning the wheel faster, so the duration of each flash became shorter. Furthermore, to obtain a bright enough flash, very strong lights had to be used, and these tended to burn the retina. At the end of the war, easy and accurate flicker was attainable by employing an electronic stroboscope which can be calibrated in fractions of a cycle per second, with a very short brilliant flash, the duration of which does not vary with the frequency.

It was found, as expected, that each flash of light evoked in the brain a characteristic electrical response. Experiments on the cortical response to light stimuli had become familiar before the war. But, now with the fresh technique, strange patterns, new and significant, emerged from the swift scribbling of the pens in all channels of the EEG. (Figure 9.)

The flash rate could be changed quickly by turning a knob and at certain frequencies the rhythmic series of flashes appeared to be breaking down some of the physiological barriers between different regions of the brain. This meant that the stimulus of the flicker received in the visual projection area of the cortex was breaking bounds; its ripples were overflowing into other areas. The consequent alteration of rhythms

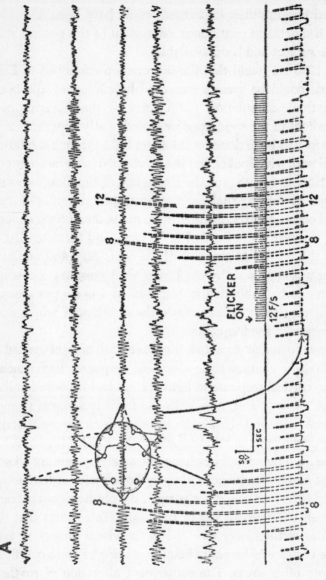

Figure 9. " . . . strange patterns, new and significant, emerged . . ." (a) The resting rhythm is at 8 c/s and flicker evokes another rhythm at 12 c/s.

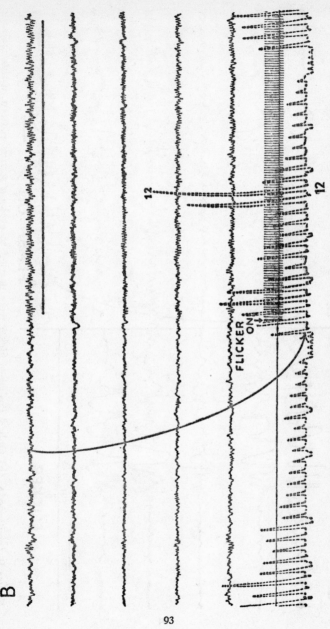

Figure 9 (continued). (b) Flicker changes the rhythms in all parts of the brain.

93

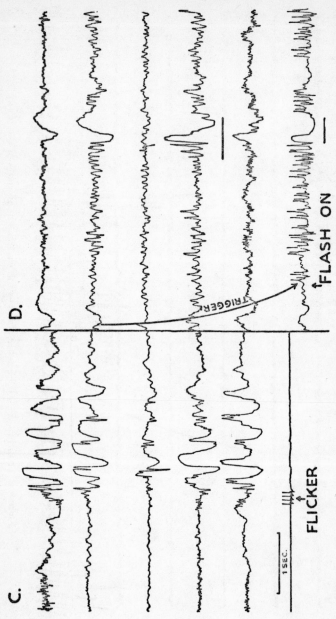

94

Figure 9 (continued). (c) Only 4 flashes of flicker were needed to evoke a seizure pattern in this epileptic. (d) Careful setting of the feedback trigger evoked a similar response in a normal subject.

in other parts of the brain could be observed from moment to moment, even by an amateur, as the red ink pen of the automatic analyser flicked its new patterns caused by the changing flicker frequencies, reporting the effect of them in one channel after another.

Flicker proved to be a key to many doors. We used it first as a clinical aid in the diagnosis of epilepsy, but from the very first its revelations were fundamental in both clinic and laboratory. Observations of many thousands of resting records, taken from epileptic patients during a quiet phase between seizures, had shown that their brain rhythms tended to be grouped in frequency bands. It was as if certain major chords constantly appeared against the trills and arpeggios of the normal activity. This harmonic grouping suggested that if a masterful conductor were introduced, the brain could be made to synchronise in a grand *tutti,* to develop under controlled conditions the majestic potentials of the convulsive seizure.

Thus it was that flicker, introduced as an aid to diagnostic interpretation, led us into new paths of research. The discovery that, in fact, appropriate frequencies of flicker did produce seizures in predisposed subjects confirmed the importance of research with this method. The new information obtained in this way about epilepsy has profoundly modified current notions about the malady.

Epilepsy has always been regarded with feelings that have varied between religious awe, professional diffidence and lay horror—with as little reason in one case as another. The specialist was unable to account for it or even adequately define it. Not that positive assertions are lacking in the literature of the subject. They can be divided into two classes: the classical—"Epilepsy is the tendency to recurring epileptic seizures"—and the romantic—"Epilepsy is a paroxysmal cerebral dysrhythmia." Those in the first class may be discounted as tautologous, while those in the second arouse the facile objection that in some cases the condition is neither paroxysmal nor cerebral, nor a dysrhythmia. The clinical researches of Penfield and Jasper in Canada and Gastaut in France are beginning to define more clearly the medical problems raised by the experimental studies.

The new information about epilepsy rather tends to excuse past failure to define or account for the malady, not because of any additional obscurity but by raising the question whether we ought not perhaps to regard it as a functional atavism rather than a degeneration. Many afflictions, regarded and treated as diseases, may be suspected of having functions of a normal physiological order. Dermatitis, inflammation of the nerve ends of the skin, is often found to have a regenerative effect and may very well have its origin in some reaction of the defence mechanisms of the organisms. A seeming mishap may have the function of averting a catastrophe. There is specific warrant, too, for speculating in this way regarding the nature of epilepsy: the similarity of a natural seizure and one induced by electric shock therapy is suggestive. And nature has many analogies. The electrical discharges of its nerve-net accompanying the movements of a jellyfish, described in an earlier chapter, are quite similar to those that occur in the human brain during the convulsive stage of a major seizure. We may be in the presence, therefore, of a vestigial echo of that first remote foreshadow of a brainstorm. On the other hand, when we have enough data for such reckoning, we may get a statistical answer indicating to what extent epileptic seizures may be a necessity for this or that degree of complexity of combination between the myriad million units of our Olympian nerve-net, considering that wherever two or three nerve cells are gathered together a seizure may occur.

This notion, that epileptic seizures are not the exclusive property of the clinically epileptic brain, is supported by observations of the effect of flicker on normal folk who have never had any kind of attack or fit. In order to compare the clinically epileptic responses with those of normals, we examined several hundred "control" subjects—schoolchildren, students, various groups of adults. In three or four per cent of these, carefully adjusted flicker evoked responses indistinguishable from those previously regarded as "diagnostic" of clinical epilepsy. When these responses appeared, the subjects would exclaim at the "strange feelings," the faintness or swimming in the head; some became unresponsive or unconscious for a few moments; in some the limbs jerked in rhythm with the flashes of light. The moment such disagreeable sensations were reported the flicker was of course turned off; recruitment of normal volunteers is not encouraged by stories of convulsions which also might quite unjustly impair the good repute of electroencephalography as a harmless experience.

The number of normal subjects who regularly and repeatedly give these epileptic responses is a small proportion of the whole population, but it is large enough to indicate that "epileptic" phenomena can be evoked in normal people by physiological stimulation of a certain type. This effect might be considered merely as a laboratory curiosity were it not that the necessary conditions occasionally occur by accident in everyday life.

Since these observations were first reported many accounts have been received from normal people who have had comparable experiences. For example, a correspondent in Amsterdam writes: "I was still in the army and my driver and I were driving home one day through an alley of trees in bright daylight. As I was tired I relaxed in my seat and closed my eyes. The sunlight that came through the trees played on my face, when suddenly I was aware that I had made some violent motions and woke up with one hand firmly on the windscreen —and this had prevented me from falling off that jeep. I was very puzzled and the next time we drove through the alley I tried the experiment all over again, but now I was all set for it. Now I could close my eyes but a very little while and I knew that if I kept on, I would lose control. When I looked straight ahead, however, it did not bother me, except that I felt a bit queer and always seemed to try to avoid the flickering light of the trees, which I did with my hands before my eyes."

In another case a man found that when he went to the cinema he would suddenly feel an irresistible impulse to strangle the person next to him; he never did actually throttle anyone, but came to himself with his hands around his neighbour's throat. This impulse was not dependent upon the subject of the film, but occurred most often if he moved his head suddenly while the film was on. When subjected to artificial flicker he developed violent jerking of the limbs when the flash rate was high—up to 50 per second—that is, about the flicker rate of the cinema projector. He could prevent the jerking by voluntary tensing of his arms. Another subject had several times "passed out" for an instant while cycling home on fine evenings down an open avenue of trees. In his case, the failure of control induced by the flicker had stopped his pedalling, and by thus slowing down he had lowered the flicker frequency and terminated the effective stimulus pattern.

In many subjects the degree of synchronisation between the flicker and the brain rhythms must be extremely precise to be effective; an error of 10 per cent may be too much. Such accuracy is difficult to maintain by hand, particularly since stimulation itself changes the frequency of the activity with which it must be synchronised. In order to keep the flicker and the brain in time, a feedback system of automatic control was adopted. This is in the form of a trigger-circuit, the flash being fired by the brain rhythms themselves at any chosen time relation with any rhythmic component of the spontaneous or evoked activity. Selection of the triggering signal can be made on a basis of amplitude or location, as well as frequency, and the delay between signal and flash can be set to anything between one second and a few milliseconds. Thus the whole gamut of relations between a great variety of flicker stimuli and any spontaneous brain-rhythm at choice can be explored.

With this instrument the effects of flicker are even more drastic than when the stimulus rate is fixed by the operator. The most significant observation is that in more than 50 per cent of young normal adult subjects, the first exposure to feedback flicker evokes transient paroxysmal discharges of the type seen so often in epileptics. This "first time" response dies away with continuous exposure, except in the 3 or 4 per cent already mentioned, suggesting that it may be connected with a specific variation in the mechanism of learning which will be described later.

Lest it should be thought that intermittent illumination, by fluorescent lights for example, is a serious hazard, it may be comforting to repeat that only a few per cent of normal people respond in this way at all, and all but very few of these only to very bright short flashes at frequencies of 10 to 20 per second. The effective flash rate of fluorescent fittings is 100 to 120 per second and the flicker is only a small proportion of the total light. Oddly enough it is not in the city, but in the jungle conditions, sunlight shining through the forest, that we run the greatest risk of flicker-fits. Perhaps, in this way, with their slowly swelling brains and their enhanced liability to breakdowns of this sort, our arboreal cousins, struck by the setting sun in the midst of a jungle caper, may have fallen from perch to plain, sadder but wiser apes.

But we must leave this subject now. The reader who wishes to know more about EEG technique and artificial stimulation as aids to the study of epilepsy, will find a lengthy account in *Electroencephalography*, published in 1950, a 438-page symposium on various aspects of the subject.

In the biological sciences it is a good principle to be your own rabbit, to experiment on yourself; in electroencephalography the practice is widespread, convenient and harmless. Whenever a new instrument is to be tested or calibrated, normal subjects from among the laboratory staff are used as "signal generators." Their brain rhythms vary very little from month to month and the new device can be tested under operational conditions in which its required performance can be predicted. When we started to use high-power electronic stroboscopes to generate flicker, with the aim of testing the hypothesis of resonant sychronisation in epilepsy, we took a large number of records from one another while looking at the brilliant flashing light. We wanted to make sure that there was no discomfort or danger and that the enormous electric current which makes the flash—a sort of artificial lightning—did not interfere with our recording instruments.

The tests were entirely satisfactory and in fact gave us much information which will be discussed later; but as well as that we all noticed a peculiar effect. It must have been observed before, but with the old-fashioned spinning-disc method of making flicker its origin did not seem so puzzling. This effect was a vivid illusion of moving patterns whenever one closed one's eyes and allowed the flicker to shine through the eyelids. The illusion is most marked when the flicker is between 8 and 25 flashes per second and takes a variety of forms. Usually it is a sort of pulsating check or mosaic, often in bright colours. At certain frequencies—around 10 per second—some subjects see whirling spirals, whirlpools, explosions, Catherine wheels.

A vivid description of the experience is given by Margiad Evans in "A Ray of Darkness": "I lay there holding the green thumbless hand of the leaf while things clicked and machinery came to life, and commands to gasp, to open and shut my eyes, reached me from across the unseen room, as though by wireless. Lights like comets dangled before me, slow at first and then gaining a fury of speed and change, whirling colour into colour, angle into angle. They were all pure ultra unearthly colours, mental colours, not deep visual ones. There was no glow in them but only activity and revolution."

With a rotating wheel generating flicker it might be argued that these impressions were due to the rapid movement of the shadow past the eye. We thought at first that a similar explanation might be adequate with the electronic stroboscope. The discharge takes place in a small spiral tube and it seemed possible that the luminous region might spin down this spiral and thus produce the illusion. But the whole brilliant flash lasts for only about twenty millionths of a second, and actually the more diffused the light, the more vivid the illusions. So we were forced to the conclusion that what we were seeing was produced, not in the light, but in the eye or in the brain.

We next supposed that our visions were derived from some peculiarity of the retina, where millions of nerve cells in matted layers transform the radiant energy of light into electrochemical changes which stimulate the optic nerve fibres. All sorts of distortions and aberrations could arise in this outpost of brain. But study of the electrical pulsations set up in the retina itself—the electroretinogram—showed no unusual effects. Furthermore we became aware of another puzzling feature of these flicker illusions: they were modified by the mental state of the subject.

These observations encouraged us to survey with greater precision the subjective illusions described by a large number of normal subjects and to analyse in detail the electrical brain responses with which they were associated. The sensation of movement and pattern, where all was still and void of detail, was surprising and contradictory. It suggested that in testing a device to study epilepsy we had stumbled on one of those natural paradoxes which are the surest sign of hidden truth.

The results obtained in these studies are too voluminous to be detailed—indeed they are being added to daily by experiment in many laboratories; but two points about records of evoked responses taken during these periods of illusion may be noticed here: 1. When the illusory sensation is primarily visual, the electrical change is in the region of the visual projection and association areas of the cortex. 2. When the sensation is related to a non-visual sensory system, the response is in the region where that system is cortically projected, the visual area response in these conditions being diminished. The experiments thus provide evidence on a grand scale and under precise control of what has already been referred to as the breaking down of some of the physiological boundaries between different regions of the brain, an overflow of visual responses into other sensory systems. It is something more than the casual "crosstalk" that causes confusion between the senses of taste and smell.

These two observations convinced us that the illusions we had first noticed ourselves, now reported with such variety by our experimental subjects, were due to some particularity of brain function. Moreover there seemed a strong possibility that many regions of the brain were involved in this deception. The visual projection area by itself is too specialised and unoriginal to create such picturesque apparitions.

We have known well enough for some time what happens in the brain at this point without knowing how it happens. We have known that the result of the final processes is that we not only see, which we can do quite absentmindedly, but that we think about what we see. We may simply enjoy the sight, or be moved by some less pleasant emotion; or it may remind us of other scenes, people, ideas; it may provoke by symbol or word some specific train of thought leading to action, or leave us satisfied or expectant. In any case, none of these results can be obtained by reception of the image on the projection area alone. Somehow this image is communicated to the other areas of the brain in which there is cognition—awareness—of sensory perceptions, or recognition of them, remembrance and association of them with other remembered sensory impressions and all their associated thoughts, feelings, ideas. We could even guess that, in order to achieve these very complex final processes, the brain must be able to communicate every item of information received in any one part of it to all its other parts. The director, in effect, does not go to the projection room; he has too many other things to direct, what with the other senses and memories and all his own reflections; rather he has some arrangement by which he is kept informed about anything that might interest him during his waking hours.

All this was asserted when our actual knowledge of the processes described was limited to the visual areas of the cortex, the existence of which could be demonstrated, and to observations in the clinic that when there was physical interference between those areas and the so-called silent areas, things were seen but not recognised. The inference from these observations was quite legitimate. It was an instance of investigation by the communication engineer's method of the Black Box: without ever looking into the box a good deal can be learned about what is going on inside by checking incoming signals against outgoing signals. For some of our problems this will be a necessary method of enquiry. But in probing further into the processes of vision, while it cannot be said that the contents of the Black Box have been completely exposed, a more direct means of enquiry was found in flicker stimulation.

The alpha rhythm—more properly, rhythms, since analysis has shown that it is nearly always compound—has had two special features of interest from quite early days. Its association with differences of personality was not at first recognised but from the very first Berger was able to show that it was affected by the opening and shutting of the eyes, and by concentration on some mental problem. In this connection it is interesting to note that the mechanism for opening the eyes is closely linked in the brain with those which maintain attention and consciousness. There are many curiosities in the relationship, however; for although closing the eyes is an essential preliminary to sleep, and may even induce sleep when the rest of the organism does not particularly require it, many people can think better—more actively, at least—with shut eyes; and visual imagery is usually more vivid without the competition of real stimuli. Tests with the eyes shut are therefore advantageous. Since the eyelids are by no means opaque, the effect of closing the eyes in a bright light is mainly to eliminate detail from vision; brilliance and colour are of course altered, but the intensity of the flicker stimulus may still be maximal with the eyes shut, using modern equipment.

Some beginning had already been made in the grouping of subjects according to the activity of their alpha rhythms in different circumstances; this was leading inevitably to the need of extending observation beyond the usual bounds of physiology into the field of physical personality. The discovery of the theta rhythm was pointing in the same direction. The adoption of flicker stimulation was itself like the projection of a new beam of light on those manifestations of brain activity. Without flicker, we could learn a good deal about EEG responses to stimulation of various kinds. But the effect of flicker was to produce in the brain, recorded in the EEG, a pattern of the same kind as the spontaneous patterns in the brain. It could also be given the same rhythmic value. That is, after ascertaining by EEG analysis the frequency of a dominant pattern in the subject, the stroboscope could be regulated to produce a flicker of the same frequency. This mirror frequency is not always the one at which flicker has the greatest effect, but with the trigger device already described, perfect synchronisation can be achieved.

Since stimulation methods, by definition, are expected to have some effect upon the subjects, note has to be taken not only of the EEG but of emotional and all other factors which may affect or be associated with the electrical changes recorded in it. The subject is asked to describe any unusual sensations; an attachment on his throat records the moment at which his remarks are made, so that the effect described can be related to the flicker frequency of the moment. The will of the subject can also be brought into play; he can, for instance, consciously and with effect resist or give way to the emotions or hallucinations engendered by the flicker, a matter of no little social interest as well as enlightenment on the question of self-discipline to be referred to later.

The greatest variety of mental experiences are described, not by any means all of them unpleasant. Some have seen profuse patterns of many colours, sometimes stable, sometimes moving; one of the first patterns we saw ourselves, the whirling spiral, recurs quite often and this may have a peculiar mechanical significance as will be suggested later. Simple sensations in other than the visual mode are experienced. Some describe feelings of swaying, of jumping, even of spinning and dizziness. Some people feel a tingling and pricking of the skin. A few subjects yielded epileptic patterns, as already described. Auditory experiences are rare; but there may be organised hallucinations, that is, complete scenes, as in dreams, involving more than one sense. All sorts of emotions are experienced: fatigue, confusion, fear, disgust, anger, pleasure. Sometimes the sense of time is lost or disturbed. One subject said that he had been "pushed sideways in time" —yesterday was at one side, instead of behind, and tomorrow was off the port bow.

Considering now these hallucinations in their simplest manifestation, how does it happen that the precise repetition of a flash, making a field of light on the retina without pattern in it, is seen as a moving pattern? The light is stationary; the eyes are shut and do not move; the head and brain are still. Yet something must move to produce moving patterns. We know that the patterns are not produced externally; they cannot reach the retina through the eye. We know that they are not formed spontaneously on the retina; they are susceptible to change by the mental state and attitude of the subject, and no anomalous effects can be detected in electrograms taken from the retina.

The imaginary patterns provoked by flicker in conjunction with alpha rhythms are produced in the brain. Their movement is the movement of some hitherto unsuspected mechanism of the brain. What is this mechanism?

The observed relations of the factors can be put in formal statements: Flicker plus x produces Movement.

$$F + x \longrightarrow M$$

Alongside this consider our other unanswered question: how is the director kept informed? How is the spatial image which is received in the projection areas transferred to the other areas for cognition? Here also is the movement of some mechanism of the brain at work. We may say that alpha plus x produces this final Image.

$$A + x \longrightarrow I$$

These are simultaneous equations in which the known factors, logically dealt with, agree with experience, and from which the unknown factors, reasonably assumed to be identical, can be eliminated by subtraction of the first equation from the second. This gives: $A \longrightarrow I - M + F$. That is, the alpha rhythm produces a motionless but intermittent image. Whatever it is that *combines* with alpha and flicker to produce a *moving* image, will combine with alpha to *transfer* an image to the final stage, a process through which of course the hallucination image also passes.

Put in this attenuated form, and given the proviso that the required mechanism has to be contained in the human head, a communications engineer would jump to the only conclusion which would seem possible to him: a scanning mechanism. The most familiar example of such mechanism is in television, where a space-pattern is most economically converted for transmission into a time sequence of impulses by the scanning mechanism of the camera. A scientist must be careful about jumping to conclusions, but if we seem to do so in this case it must be said that it is several years since we took off with the first surmise, and that the American psychiatrist and psychologist, Warren McCulloch, landed about the same time, quite independently, on the same hypothetical spot, and we have not been warned off by the authorities.

There is much to be said for the hypothesis. It is necessary to assume some mechanical principle which could be incorporated in apparatus compact enough to be carried about in the head, and no other principle of this character is known. Also, both equations are satisfied by $x =$ scanning mechanism. The hypothesis indeed was not reached dialectically, but as a result of long scrutiny of the behaviour of the alpha rhythms under both conditions—that is, with and without flicker.

There was, for instance, the curious coincidence between the frequency of the alpha rhythms and the period of visual persistence. This can be shown by trying how many words can be read in ten seconds. It will be found that the number is about one hundred—that is, ten per second, the average frequency of the alpha rhythms. Scanning also explains an optical illusion which can be tested. If two lights are available in the same field of vision, when one of them is turned off just before the other is turned on it looks as if there was a movement of light from the first to the second. The illusion is accounted for by the scanning interval.

The cessation of the alpha rhythms in most subjects when the eyes are opened and a pattern is seen, or when one is induced by suggestion of an image of a visual or mental kind, is also significant. It suggests that the alpha rhythms are a process of scanning—searching for a pattern—which relaxes when a pattern is found. It is as if you were looking for one particular word in a page of print; you scan the page, line by line, until you come to the word; then the scanning movement ceases; anyone watching your eyes could tell when you had found what you were looking for.

The manner in which such a system might work in the brain is illustrated in a device which we had developed for quite another purpose, for re-converting line records, such as EEG's, back into electrical changes. (See Figure 10.) A cathode-ray oscilloscope, photo-electric cell and amplifier are arranged in a feedback system such that the photo-cell "looks"

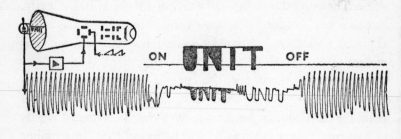

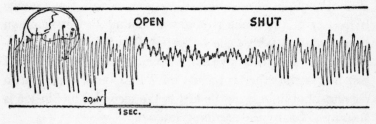

Figure 10. ". . . the alpha rhythms may be a process of scanning . . ." A Profile Scanning Device. When a feedback scanning system has nothing to scan it tends to oscillate. When it scans a pattern, its movements reproduce a facsimile of the pattern on a time base. Below, the alpha rhythms of a normal subject seem to suggest similarity of mechanism in model and brain.

at the light spot on the screen of the oscilloscope, the resulting voltage being amplified and applied to the Y-plates of the oscilloscope, thus deflecting the spot. The deflection reduces the apparent brightness of the spot, thus altering the voltage, and giving the spot a tendency to hover at a point near the edge of the screen. If an opaque object is placed over the part of the screen toward which the spot tends to move, the spot cannot get behind it since its own disappearance abolishes the force deflecting it. If a linear scanning voltage is applied, the spot will slide along the profile of any curve placed on the screen. If, however, in certain conditions, there is no pattern for it to scan, the system tends to oscillate in a manner analogous to alpha rhythms. The oscilloscope screen may represent the projection area of the visual cortex, which may either be vacant or occupied by the projection upon it of an electroneural chart of the visual field. The electron beam and spot of light represents the fluctuating electrical activity generated by a contiguous association area through excitation of chains of neurones leading through the projection area. The photocell represents the other end of the neurone chain which receives stimuli back from the projection area and completes the retroactive loop.

The way in which the pattern of alpha rhythms is altered or suppressed by imposed mental or visual patterns can be followed in EEG recording; it can be watched with greater dramatic effect in the toposcope where the dominant pattern can be seen taking possession of the visual projection and association areas. (See Figure 6.) When flicker is used, the display given by the toposcope comes near to being a moving picture of a mind possessed in quite another way. The correspondence between the extent and complexity of the evoked responses on the one hand, and the hallucinations of the subject on the other, is striking. The more vivid and bizarre the experience of the subject, the farther from the visual areas are the evoked responses, and the more peculiar their form and geometry. The behaviour of the spontaneous and artificial rhythms in these conditions is strictly in accordance with the effects to be expected from a scanning mechanism. The interference of flicker with a normal process of scanning would be shown on a television screen by illuminating the television studio with a flickering light. The effect of this on the picture would be most unpleasant, indeed hard to bear; blobs of light would dart giddily about the screen. Similar confusion in the brain is seen in our records and toposcopic observations, when the conflict between the two time patterns, the inherent scanning rhythms of the brain, and the flicker, produce a brain storm as wild as any distortion on the television screen.

A general theory such as this, about an organ of such complexity and variability as the brain, cannot be verified or established by any single experiment, as would be accordant with the principles of classical physiology. It will be a matter of day-to-day experience in many laboratories, to learn in time whether the hypothesis will hold, as one new phenomenon after another comes to test its suppositions. The time may not be long, however; the new information being obtained by new techniques and apparatus is voluminous and overflowing with intriguing suggestions. For instance, the whirling spiral so many people see under flicker, is it not perhaps an indication of the very path taken by the scanning point in the pattern it makes every tenth of a second?

It would be tempting to develop here a complete theory of scanning as the final stage of all the sensory processes; for it seems to serve equally well to account for the cognition of other than visual sensations. We shall be seeing later, for instance, how appropriate it is for interpreting the theta rhythm as a scanning for pleasure, a supposition well supported by the mental experiences accompanying its evocation. But now some mechanical pets are waiting in the next chapter, prepared to explain in their simple fashion some of the phenomena already discussed, or, like the White Knight's inventions, divert us until we reach the Eighth Square of thought and personality.

CHAPTER 5

Totems, Toys and Tools

Am Ende hängen wir doch ab
Von Creaturen die wir machten.
                                        Goethe

THE MAKING of images may have a variety of
purposes, and the intention of the maker is not always clear
to those who look on them. This chapter will be about mimicry
of life, so it may be well to explain at the outset why a scientist
resorts to methods which might seem more appropriate to the
entertainer, the artist or the priest. The suspicion that the
scientist is not quite sincere in professing that his purpose is
purely mechanical and illustrative goes a long way back. The
notion of magic is deep-rooted. A term fairly applicable to
our subject, for instance, "Electro-biology," is found in Roget's
Thesaurus (authorised copyright edition, 1946) listed under
"Acts of Religion," in the sub-section, "Sorcery."

It all goes back to the totems of primitive man, to the images
we used to make of game or enemy on the walls of the cave,
to the images of their rivals into which village boys and girls
still hopefully stick pins. Then the totem is set up for worship, developing perhaps into forms and images of great
beauty and unchallengeable piety and found in greater or less
profusion as iconoclast alternates with iconolater. There are
images in the canon of all faiths, excepting that of the Semites,

who discovered the unrepresentable behind their graven images and abolished them, and that of the Quakers and Christian Scientists, who never had any. We are daily reminded how readily living and even divine properties are projected into inanimate things by hopeful but bewildered men and women; and the scientist cannot escape the suspicion that his projections may be psychologically the substitutes and manifestations of his own hope and bewilderment.

There is, however, a well-defined difference between the magical and the scientific imitation of life. The former copies external appearances; the latter is concerned with performance and behaviour. Until the scientific era, what seemed most alive to people was what most *looked* like a living being. The vitality accorded to an object was a function primarily of its form. Even ships, most wilful and exacting of all contrivances, were not fully alive until they had been given a figurehead and were formally christened and blessed, the figurehead being an embodiment of the ship's soul.

Moreover, any performance by a sacred image must be magical—a Virgin with an ingenious mechanical smile would get no worship. Again, an image may be only a toy. The technical genius of the Swiss watchmakers was really wasted on their delicate clockwork automata; they arouse only a passing interest because they are neither sacred nor, like life, unpredictable, their performance being limited to a planned series of motions, be it a boy actually writing a letter or a girl playing a real keyboard instrument.

With the coming of steam, and later electricity, a new sort of automatic device became necessary, not as totem or toy but as tool, something to enable a machine to control its own effective use of the power it generates. The first steam engine, left to itself, was unstable—pressure went down when power was used and boiler blew up when it was not used. Watt introduced the safety valve and automatic governor which stabilise by themselves both boiler pressure and engine speed. These two important devices were taken rather as a matter of course by engineers, but the great Clerk Maxwell devoted a paper to the analysis of Watt's governor. Maxwell was perhaps the first to realise the significance of this key process of feedback. Later, physiologists pointed out that Watt had incidentally constructed in principle the first working model of a reflex circuit similar to what they describe in the organisation of sense organs, nerves and muscles. Self-regulating devices, so common today that their existence is taken as a matter of course by the millions who use them, are the gas oven and refrigerator thermostats and the automatic volume-control in your radio which maintains the volume of sound within certain limits.

But until the invention of the thermionic tube, any ambition to construct a true imitation of a reflex circuit, let alone follow in the footsteps of Mary Shelley and her Frankenstein, would have been a scientific vanity. Consider for a moment what a "monster" indeed would be a working model of the brain, even with the miniature tubes and other components available today.

First, some model living cells—what would be the smallest size theoretically possible? They would have to generate, accumulate, discharge and regenerate their own minute voltages. If the necessary chemical contents of one cell, with insulation, capacitance and trigger mechanism, could be crowded into a quarter of a cubic inch, it would be a miracle of construction, though still gross compared with the original. To put aside the standard number of these cells, 10,000 million, for one brain, while considering the wiring problem, warehousing space of about one and a half million cubic feet would be required.

The kind of circuit necessary to imitate the behaviour of a nerve fibre was a laboratory problem of no mean order and was only recently solved. Just consider the specification for a model nerve. It must conduct an impulse in both directions, wherever the impulse is applied; it must convey what is known as an all-or-nothing impulse, followed by a refractory period—when no impulse is carried; moreover, the travelling impulse must be self-propagating and non-decremental—that is, it must travel any distance without decreasing its volume or voltage. The difficulties of design and construction were overcome only after many experiments; a diagram of the circuit will be found in Appendix A. Supposing a sufficient number of these simplest of imitation nerves were constructed, more millions of cubic feet of warehousing would be required.

At this point we might have to consider cost. A cell with one fibre might conceivably be made for about a dime,—in all, say, $1,000,000,000. Wiring connections, 10^{20} of them, at about two cents each, say $2,000,000,000,000,000,000. Power required would be at least a million kilowatts, even if the transistor crystal were used instead of the prodigal thermionic tube; the human brain runs on 25 watts. The cost would be incalculable. And the comparable cost of producing a first-class living brain? Twenty thousand dollars?

No, nobody is going to make an artificial brain on those lines, even with the smallest and cheapest conceivable midget parts. An entirely different approach seemed necessary to make it a practical problem, if we were to learn about life by imitation as well as observation of living things.

One other possible approach to the baffling complexities of the nervous system there did seem to be. What we have just reduced to absurdity is any prospect of reproducing all its elaboration of units in a working model. If the secret of the brain's elaborate performance lies there, in the number of its units, that would be indeed the only road, and that road would be closed. But since our enquiry is above all things a question of performance, it seemed reasonable to try an approach in which the first consideration would be the principles and character of the whole apparatus in operation.

This raised a new question in brain physiology. It meant asking whether the elaboration of cerebral functions may possibly derive not so much from the number of its units, as from *the richness of their interconnection*.

As a hypothesis, this speculation had the great advantage that its validity could be tested experimentally. An imitation of two or three interconnected elements, including reflex circuits to demonstrate their behaviour, should be a simple matter for a laboratory that had produced the EEG analyser and the toposcope.

But how many elements would be needed to demonstrate this? How many elements could be managed? Here we ran into an unforeseen difficulty—that of reckoning how many ways of behaviour are possible with a given number of elements. Neither mathematicians nor communications engineers were acquainted with precisely the problem this raised.

Reduced to its simplest form the question is, How many ways of behaviour would be possible for a creature with a brain having only two cells? Behaviour would depend on the

activity of one or both of these cells—call them A and B. If
(1) neither is active, there would be no action to be observed;
if (2) A is active, behaviour of type *a* would be observed; if
(3) B is active, behaviour of type *b;* if (4) A and B are both
active, but independently, there would be behaviour of both
types *a* and *b,* mixed together; if (5) A is "driving" B, type *b*
would be observed but subordinate to A; if (6) B is "driving"
A, type *a* would be subordinate to B; if (7) A and B are "driving" each other, behaviour will alternate between type *a* and
type *b.* The internal states of such a system in these seven
modes may be represented symbolically as:

$$O, A, B, A + B, A \rightarrow B, A \leftarrow B, A \leftrightarrows B$$

with behaviour types:

$$o, a, b, a + b, b(fA), a(fB), ababab \ldots$$

From the above it will be seen that the first four ways of
behaviour would be identifiable by simple observation, without interfering with the system, whereas the last three could
only be identified by operating on the system—by, as it were,
dissecting out the arrows.

Thus, with only two elements interconnecting, there are
seven modes of existence. With six units there would be
enough modes to provide a new experience every tenth of a
second throughout a long lifetime. If there are *n* elements,
the number of modes is given approximately by $M = 2^{(n^2 - n)}$
for six or more elements. So an extremely varied behaviour is
possible when a modest collection of elements is capable of
intricate interaction. If the reader would like to calculate how
many possible modes there might be in the brain, with its
ten thousand million elements, take the number 2 and double

it a hundred million million million times—which is making a fantastic "monster" of the brain itself.

How many functional units, then, we may ask, must a brain contain in order to account for its behaviour, assuming the interconnection may be as complete and varied as it is usually supposed to be? As a start, we may suggest that the number of active elements may be of the order of 1,000 ensembles of homologous neurones. If these are capable of dynamic combinations and permutations according to the above equation, then the number of behaviour patterns possible would be of the order of $10^{300,000}$—that is, if written out at length, a figure so long that it would fill more than a hundred pages of this book. Even were many millions of permutations excluded as being lethal or ineffective, the number is still large enough to satisfy the requirements of individuality and plasticity.

The outlook for an experimental model brightened at once with the problem reduced to the behaviour of two or three elements. Instead of dreaming about an impossible "monster," some elementary experience of the actual working of two or three brain units might be gained by constructing a working model in those very limited but attainable proportions.

But if the performance of a model is to be demonstrably a fair imitation of cerebral activity, the conditions of stimulation and behaviour must equally be comparable with those of the brain. Not in looks, but in action, the model must resemble an animal. Therefore it must have these or some measure of these attributes: exploration, curiosity, free-will in the sense of unpredictability, goal-seeking, self-regulation, avoidance of dilemmas, foresight, memory, learning, forgetting, association of ideas, form recognition, and the elements of social accommodation. Such is life.

This rules out the charming creations of the Swiss watchmakers; in spite of the intention of the artificers, evident in the name of "automates," they are not automata in the sense of being endued with spontaneous motion, and in no sense of the word have they any claim to autonomy or self-regulation. Their performance is prescribed or, as we say today, programmed from beginning to end of the starting impulse.

Nor can variety of programming endow a machine with the autonomous qualities of a true mimicry of life. The computing machine is a model of Nineteenth Century predestination; it was devised more than a hundred years ago without a thought of imitating the living brain. The works of two mathematicians of the day, Boole and Babbage, taken together, provide all the basic theoretical and mechanical knowledge necessary for a blueprint of one of the giant machines, lacking only the thermionic tube to give it the speed of an electron instead of a piston. A third famous Victorian mathematician immortalised their manner of operation for all children, young and old: "Can you do addition?" the White Queen asked. "What's one and one and one and one and one and one and one and one and one and one?" "I don't know," said Alice. "I lost count." The algebra of Boole is the algebra of yes and no. Computers are essentially machines that do sums in that fashion but don't lose count. Babbage made one in 1822; a few years later when he was constructing a much more elaborate one for the Government he discovered a new mechanical arrangement which he called "the engine eating its own tail." The discovery no doubt made it very tedious to go on with the machine already begun, on which he had spent £6,000 of his own and £17,000 government money; in any case official support was withdrawn in 1842, and the completed fragment—"a beautiful machine which does its work with unerring accuracy but is useless," as described not long after Babbage died in 1871—ended in the South Kensington Museum. His analytical engine was never constructed. It is recorded that "in his later years he was chiefly known by his fierce hostility to organ-grinders"—envious perhaps of the persistent demonstration of their "engines" while his remained on paper, 400 detailed drawings and many volumes of notes. It was, in the modern style, to be programmed by punched cards, with special cards for particular functions; it was to calculate the numerical values of any formula or function of which the mathematician could indicate the method of solution, relieving him "of all the drudgery of computing," and was of course to print its answer. Among the wonders of constructional anticipation it is comparable with those of Hero of Alexander and Leonardo da Vinci.

The first attempt to make a machine that would imitate a living creature in performance, as distinguished from appearance, seems to have been suggested by the familiar test of animal intelligence in finding the way out of a maze. Thomas Ross in 1938 made a machine in America which successfully imitated this experiment. By trial and error it could "learn" to find its way to a correct goal on a system of toy train tracks. Another tram-like creature of the same species, *Machina labyrinthea,* to give it a mock-biological name, was also built by an American, R. A. Wallace, in 1952, "to demonstrate that a relatively small and simple digital computing machine can solve a class of conceptual problems other than numerical calculations." It also runs on minute rails, with automatic switches, or "choice points," of which it can cope with 63 and get home without programming as to the order of exploration or any assistance in summating its experience. Once it has found its way home, its choices are pre-set—it has programmed its own route and, restarted from the same point, will go direct home without error. Claude Shannon has also devised a maze-learning creature, less rail-bound, a sort of electro-mechanical mouse that fidgets its way out of confinement.

Thus, within its limits, with a predestined end and a predictable course, *M. labyrinthea* is goal-seeking and self-regulating. Like other machines, it is said to have a kind of memory. But the use of the word at this juncture is unfortunate, seeming to claim a likeness to the flesh that cannot be upheld in detail. Storage of information there certainly is, but it is as unequivocal as the information stored in a book and, being metallic, less perishable. Human memory is subtle, variable, reinforceable in strange roundabout ways, and fallible. The engineer does not imitate it and does not wish to. "Memory" in this connection had better be forgotten with "giant brains."

Another machine with a predestined end, but quite unpredictable behaviour in reaching it, is Ashby's Homeostat. This creature, *Machina sopora*, it might be called, is like a fireside cat or dog which only stirs when disturbed, and then methodically finds a comfortable position and goes to sleep again. There are a number of electronic circuits similar to the reflex arcs in the spinal cord of an animal. They are so combined with a number of thermionic tubes and relays that out of 360,000 possible connections the machine will automatically find one that leads to a condition of dynamic internal stability. That is, after several trials and errors, the instrument, without any prompting or programming, establishes connections which tend to neutralise any change that the experimenter tries to impose from outside. So far as it goes, the homeostat is a perfect example of self-regulation by negative feedback—in fact, it is all negative feedback, like a steam engine that works nothing but the safety valve and governor. A very curious and impressive fact about it, however, is that, although the machine is man-made, the experimenter cannot tell at any moment exactly what the machine's circuit is without "killing" it and dissecting out the "nervous system"—that is, switching off the current and tracing out the wires to the relays.

What we find in this fireside companion is not only the virtue of self-control and the blessing of homeostasis, not only an exemplification of placidity, but also of plasticity, one of the basic principles that seem to govern animal engineering. This means roughly that every part of the mechanism is reversible, interchangeable and expendable—but not replaceable. As a description of a system of interconnections, this is really only half true about the brain, or only true about half the substance of the brain. Overlooking this important point of internal resemblance to an animal nervous system, and judging *M. sopora* entirely by its behaviour, the naturalist would classify it as a plant. More recently Ashby has made and studied a simple machine, functionally related to *M. labyrinthea,* which adjusts its internal connections when presented with two successive stimuli. He has also begotten an heir to Homeostat—with 25 times as many elements.

We now come to an electro-mechanical creature which behaves so much like an animal that it has been known to drive a not usually timid lady upstairs to lock herself in her bedroom, an interesting blend of magic and science.

The first notion of constructing a free goal-seeking mechanism goes back to a wartime talk with the psychologist, Kenneth Craik, whose untimely death was one of the greatest losses Cambridge has suffered in years. When he was engaged on a war job for the Government, he came to get the help of our automatic analyser with some very complicated curves he had obtained, curves relating to the aiming errors of air gunners. Goal-seeking missiles were literally much in the air in those days; so, in our minds, were scanning mechanisms. Long before the home study was turned into a workshop, the two ideas, goal-seeking and scanning, had combined as the essential mechanical conception of a working model that would behave like a very simple animal. At the same time, this conception held promise of demonstrating, or at least testing the validity of, the theory that multiplicity of units is not so much responsible for the elaboration of cerebral functions, as the richness of their interconnection. With the minimum two elements there should be seven modes of existence. And there was another good reason, apart from the avoidance of unnecessary mechanical complications, for the utmost economy of design in *Machina speculatrix*, inevitable name of the species for the discerning, though "tortoise" to the profane; it would demonstrate the first of several principles exemplified in the mechanisms of most living creatures. A few notes on these principles will illustrate the behaviour of *M. speculatrix*, from the observation of which indeed much has been learned about them.

1. *Parsimony.* The Nineteenth Century raised its poetic

eyebrows at "Nature's prodigality"; the Twentieth Century is no less surprised by the economy of structure and function discovered in the mechanics of life. There are very few redundant organs in present-day animals, and many parts of the body were originally something quite different. "Mend-and-make-do" is a popular slogan in the struggle for existence. In *M. speculatrix* the number of units corresponding to nerve cells is limited to two; there are two miniature tubes, two relays, two condensers, two small electric motors, two batteries. These two "sense reflexes" operate from two "receptors" —one a photo-electric cell, which gives the organism sensitivity to light, the other an electrical contact serving as a touch receptor, which gives it responsiveness to material obstacles. The variations of behaviour patterns exhibited even with such economy of structure are complex and unpredictable.

2. *Speculation.* A typical animal propensity is to explore the environment rather than to wait passively for something to happen. This faculty gives the device its name and distinguishes it from other machines. The most elaborate computing machine does not look round for problems to solve; nor do *M. labyrinthea* and *sopora*. But *M. speculatrix* is never still except when "feeding"—that is, when the batteries are being recharged. Like the restless creatures in a drop of pond water, it bustles around in a series of swooping curves, so that in an hour it will investigate several hundred square feet of ground. In its exploration of any ordinary room it inevitably encounters many obstacles; but, apart from stairs and fur rugs, there are few situations from which it cannot extricate itself.

3. *Positive tropism.* Sensory susceptibility to the attrac-

tions of the environment. The one positive tropism of *M. speculatrix* is exhibited by its movement towards lights of moderate intensity. The photo-cell, amplifier and motors are connected in such a way that, when an adequate light signal is received, the exploratory behaviour is checked and the organism orientates itself towards the light and approaches it. Until it "sees" a light, the photo-receptor is in continuous rotation, scanning the horizon for light signals. This scanning process is linked with the steering mechanism in such a way that the "eye" is always looking in the direction of movement; thus, when a signal is received, from any direction, the machine is in a position to respond without too much manoeuvring.

4. *Negative tropism.* Certain perceptible variables, such as very bright lights, material obstacles and steep gradients, are repellent to *M. speculatrix;* in other words, it shows negative tropism towards these stimuli. Observing the principle of parsimony, this is accomplished without introduction of additional components, by making any slight displacement of the organism's shell close a contact which converts the photo amplifier into an oscillator; this causes alternating movements of butting and withdrawal; so that the robot pushes small obstacles out of the way, goes round heavy ones, and avoids slopes. This device automatically introduces the next important principle.

5. *Discernment.* Distinction between effective and ineffective behaviour. When the machine is moving towards an attractive light and meets an obstacle, or finds the way too steep, the induction of internal oscillation does not merely provide a means of escape—it also eliminates the attractiveness of the light, which has no interest for the machine until

after the obstacle has been dealt with. There is a brief "memory" of the obstacle, so that the search for lights, and attraction to them when found, is not resumed for a second or so after a material conflict.

6. *Optima.* A tendency to seek conditions with moderate and most favourable properties, rather than maxima. The circuit of *M. speculatrix* (shown in Appendix B) is so adjusted that exploration is undertaken in darkness and moderate lights are attractive, whereas bright lights are repulsive. Thus the machine can avoid the fate of the moth in the candle. Also, with the scanning device, it can avoid the dilemma of Dante's free man, *intra due cibi,* or of Buridan's ass, which starved to death, as some animals acting tropistically in fact do, because two exactly equal piles of hay were precisely the same distance away. If placed equidistant from two equal lights, *M. speculatrix* will not aim itself half-way between them, but will visit first one and then the other.

7. *Self-recognition.* The machines are fitted with a small flash-lamp bulb in the head which is turned off automatically whenever the photo-cell receives an adequate light signal. When a mirror or white surface is encountered the reflected light from the head-lamp is sufficient to operate the circuit controlling the robot's response to light, so that the machine makes for its own reflection; but as it does so, the light is extinguished, which means that the stimulus is cut off—but removal of the stimulus restores the light, which is again seen as a stimulus, and so on. The creature therefore lingers before a mirror, flickering, twittering and jigging like a clumsy Narcissus. The behaviour of a creature thus engaged with it own reflection is quite specific, and on a purely empirical basis, if it were observed in an animal, might be accepted as evidence

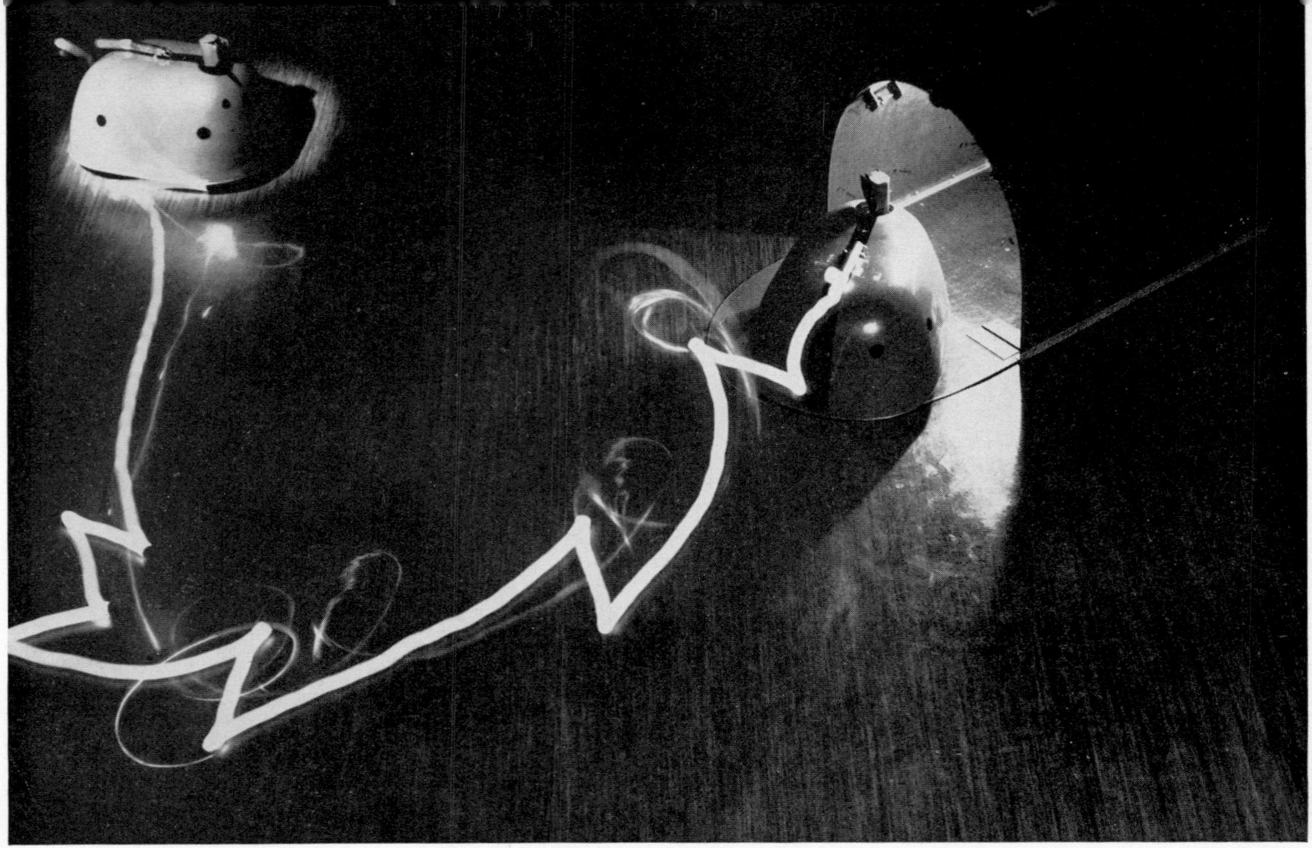

Figure 11 " ... moderation gives place to appetite." *Speculatrix* finds her way home.

of some degree of self-awareness. In this way the machine is superior to many quite "high" animals who usually treat their reflection as if it were another animal, if they accept it at all. This leads on to:

8. *Mutual recognition.* Two creatures of the same type, attracted by one another's light, both extinguish the source of attraction in themselves in the act of seeking it in others. Therefore, when no other attraction is presented, a number of the machines cannot escape from one another; but nor can they ever consummate their "desire," and when seen from the back or side a fellow creature is merely an obstacle. In a sense, then, a population of machines forms a sort of community, with a special code of behaviour. When an external stimulus is applied to all members of such a community, they will of course see it independently and the community will break up; then, the more individuals there are, the smaller the chance of any one achieving its goal, for each individual finds in the others converging obstacles.

9. *Internal stability.* One of the advantages of making a moderate light a positive stimulus is that this can be used as a sign or symbol for the energy which the creatures require for their sustenance—electricity. A light is placed in their "hutch" is such a position that they are attracted to it, and therefore tend to enter the hutch of their own accord. However, if their batteries are fully charged, the intensity of the light operates the repelling circuit when over the threshold, and they withdraw for further exploration. When their batteries *require* recharging, on the other hand, moderation gives place to appetite and the light continues to exert an attraction until they are well within their quarters. (See Figure 11.) At this point, contacts on the side of the shell

can engage with others in the hutch, thus closing the battery-charging circuit. Current flowing in this circuit operates a relay which turns off the power to their sensory and motor systems, so that the machine remains motionless until, the battery voltage having risen, the charging current falls and their internal mechanism is once again energised. This arrangement is very far from perfect; there is no doubt that, if left to themselves, a majority of the creatures would perish by the wayside, their supplies of energy exhausted in the search for significant illumination or in conflict with immovable obstacles or insatiable fellow creatures.

Some of these patterns of performance were calculable, though only as types of behaviour, in advance; some were quite unforeseen. The faculties of self-recognition and mutual recognition were obtained accidentally, since the pilot-light was inserted originally simply to indicate when the steering-servo was in operation. It may be objected that they are only "tricks," but the behaviour in these modes is such that, were the models real animals, a biologist could quite legitimately claim it as evidence of true recognition of self and of others as a class. The important feature of the effect is the establishment of a feedback loop in which the environment is a component. This again illustrates an important general principle in the study of animal behaviour—that any psychological or ecological situation in which such a reflexive mechanism exists, may result in behaviour which will seem, at least, to suggest self-consciousness or social consciousness.

The way in which the social behaviour of the models breaks down under the influence of a competitive struggle for a common goal, imitates almost embarrassingly some of the less attractive features of animal and human society. That is the fault of the maker, who could, however, easily endow his creatures, as man is endowed, with a discriminatory recognition circuit that would function in an emergency, a "women-and-children-first" reflex.

Further, it would only be a matter of patience and ingenuity to endow *M. Speculatrix* with other "senses" besides sight and touch, to enable it to respond to audible signals audibly, and so forth; also to provide it with hands—with a different tool for each finger, dream of our electronic childhood! For the time being we have removed, and shall presently be describing as characteristic of a separate species, *M. docilis*, that part of the brain with which it could learn how to use any tools put into those hands. There is no serious difficulty about the elaboration of function, once the principles of mechanical "life" have been demonstrated in a working model. If the principles are preserved, no matter how elaborate the functions of the machine, its mimicry of life will be valid and illuminating. On the other hand, if the principles are abandoned in favour of programming the machine for special purposes, as Wiener foresees, the result may be productive, the machine may entirely supplant human labour in the factory, but it will be of little interest to the physiologist. It will no longer be part of a mirror for the brain.

The character of *M. speculatrix* as the prototype of an electro-mechanical species, however, is not dependent on the possibilities of elaboration, but contrariwise on the impossibility of any simplification of the functional mechanism. Little virtue or interest would lie in achieving lifelike effects with a multiplication of mechanisms greater than life would tolerate. Creatures with superfluous organs do not survive; the true measure of reality is a minimum. Occam's razor is as sharp in the struggle for existence as it is in wordy strife.

This law can be seen at work in the progeny of *M. speculatrix:* the sports and unadaptable mutations fade away, the successful imitations already form more than one sub-species. Most interesting among these is an acknowledged offspring reported last year by a brilliant young American engineer, Edmund C. Berkeley. It is, not inappropriately, a squirrel compared with the British tortoise; "a squirrel gathering nuts" was the specification; meanwhile it is practising on golf balls with its ingenious scoops. It would not in any case live on nuts, even if it could gather them; it subsists on the same fare as its prototype, on the design of which also its essential mechanisms are based. Its creator calls it Squee; in the annals of mock-biology it is likely to be remembered as *M. speculatrix berkeleyi.*

The educated laboratory mascot, *M. docilis,* the "easily taught" machine, will presently be found worthy of separate attention. But the untutored ones are not to be despised. As toys they refresh the spirit of the laboratory children we all are, leading us to familiarity with more and more elaborate mechanisms. As tools they are trustworthy instruments of exploration and frequent unexpected enlightenment. As totems they foster reverence for the life they have so laboriously been made to mime in such very humble fashion—and still would foster it even should they, creatures of "sorcery" peering into the dim "electro-biological" future in search of a *deus ex machina,* look up at us and declare that God is a physiologist.

CHAPTER 6

Learning about Learning

Time and Education begets Experience; Experience begets Memory; Memory begets Judgement and Fancy; Judgement begets the strength and structure, and Fancy begets the ornaments of a Poem.

Hobbes

THE NEW techniques with which to fashion a mirror for the brain have already provided glimpses of the mechanisms of its more accessible parts, where the reception of sensory impressions can be tested with controlled stimuli; and some of its more simple functions have been imitated. This has suggested various physiological theories, including:

1. A theory of internal scanning as a function of electrical brain rhythms in the final stage of sensory perception.

2. A theory that the elaboration of brain function is due to the richness of interconnection, rather than the multiplicity, of its parts.

3. The outline of a general theory of organic construction.

These theories were reached in procedures such as the father of modern physiology, Claude Bernard, desired for a "true scientist," a methodical feedback system of "experimental theory and experimental practice," the one always leading back to the other. It seems timely to recall this principle when setting now a course new to physiology.

The subject of learning was naturally reserved to those who studied the mind, without reference to the mechanisms of the brain, while physiologists neglected that organ. Bernard is invoked for the assurance that scientific method will not become less methodical in his sense when the field widens and complications increase; it will continually alternate between practice and theory, devising experiments and apparatus appropriate to the complexities of the matter. In this chapter, after various ways of learning have been discussed, two converging paths of research, stemming respectively from the initiative of Pavlov and Berger, will be traced. This will lead in the next chapter to a description of the processes of associative thinking and the mechanism of learning, illustrated by a working model.

A dictionary definition of learning is: any acquisition by an organism of knowledge, skill, or modes of behaviour. The psychological literature of learning was already vast long before anyone thought of looking for living mechanisms of learning; nor is it the fault of the psychologists if the brain is still neglected by them; the new techniques are still too complicated. On the other hand, some of the old investigations, upon which much psychological doctrine is based, are too simple for this world.

The classical experiment of Wertheimer, for instance, hero of Gestalt, still a popular doctrine in educational circles, was the following: Chicken feed was placed in two bowls, one light and the other medium grey; both were accessible to a hen, but she was always driven away from the medium grey one, so acquired the habit of feeding always from the light grey one; when, in place of this light one, a dark grey bowl was offered along with the medium, what was the delighted surprise of Herr Wertheimer when the hen chose to eat from the medium bowl! "The dog it was that died." Why was the experimenter surprised? Did he expect the hen to remember the precise shade of grey that was taboo? According to his followers, the hen's ability to distinguish the lighter shade of two is not associational but somehow a confutation of associational ideas of learning. A pathetic fallacy. What the nervous system receives from the sense organs is information about differences—about ratios between stimuli.

Professor Thorndyke's cats have been almost as influential as Wertheimer's hen. They get out of a box by a wild scrambling, which is modified by experience; and some psychologists think that adult learning may be of that order. They find it difficult to believe that Köhler's apes, in another famous series of tests, associate stick and fruit before using the one to get the other. They prefer to say that "the situation becomes organised," and that the deed is done in a flash of illumination. On the other hand, the doctrine of associational learning runs to two antagonistic extremes of behaviouristic theory, in Russia and in America.

Too many of these ideas about mental functions are not in the field of "experimental theory and experimental practice"; they are closed theories, conclusions; their proponents or promoters, temperamentally intolerant of uncertainty, are so eager for a single theory applicable to all types of learning that anomalous facts have to be crammed into the bag with whatever statistical wrapping can be found for them. The truth seems to be that there are many ways of learning and that the field is open for a physiological description of some of them.

Casting about for the simplest way of learning, we meet

at once the old question of what is meant by instinct, a word so widely abused that it is often deliberately avoided and some more specific term used in its place. Can it be said that what is done "instinctively" has been learnt?

Some enlightening results have come from studies of the behaviour of newborn animals, especially birds, which indicate the existence of a hitherto quite unsuspected mechanism. Lorenz first made known the curious fact that whatever a gosling, for instance, first sees in the hours after hatching, be it bird, beast or man, the gosling will follow as it normally would follow its mother. The fixation is demonstrated as not essentially on the mother but on any first moving object perceived. There is no specific mother or child instinct; it is literally blind instinct, so far as choice of the love-object is concerned. It is an ever fixèd mark. How persistent is the fixation was shown in another experiment in which the first thing presented to the eyes of a budgereegah remained for ever its only object of attachment and its days of courtship were spent in trying to make love to a ping-pong ball.

This miniature tragedy might seem to support the Freudian choice of sex, rather than love, as the basic impulse. But we realise today that this was an unfortunate narrowing of the truth. It is quite clear that this "learning by imprinting," as it is called, is not related to sex development; on the contrary, it is exclusively tied to the earliest phase in the life of the animal. If the chick is not allowed to see any moving object at all in its first hours, and only later on is presented with one, no fixation, no imprinting, takes place.

A working model of the mechanism is simple. A feedback trigger circuit provides channels for the reception of a number of possible stimuli, and is so constructed that, when any one of them is activated, it locks on, and a common reflex puts all the other channels out of action—that is, excepting the one already locked.

Mammals seem to be less susceptible to mental imprinting than birds, though it has been observed in sheep with a social implication. If a lamb is removed not only from its mother but from all other sheep for the first ten days after birth, it loses its sheeplike character of following the flock and does not develop normal ties with it. The missing object here is not a mother figure, though lambs in an early phase seem to be as sensitive to the lack of this as the human child is.

Some investigators also believe they have found the key to a baby's first smile in these impersonal "innate releaser mechanisms." Spitz found that a mask with two eyes and a nose waved slowly to and fro on the end of a broomstick was enough to elicit the first smile, between the tenth and twentieth weeks of life, all smiles before that period being attributable to wind. But it is at least a permissible question whether this is "imprint" or association. It would be strange if, in ten weeks, the appearance of a face had not already become associated with smiling fortune in the baby's mind. For the human brain, even at birth, is so highly organised, the electrical rhythms that sweep it are so suggestive of searching mental activity, that it is difficult amidst so much complexity to tell how soon the ape is left behind. The good fairy's gift of learning by association is found in every cradle.

We know that a new-born child can do three things, it can cry and suck and has already—shall we say, learnt to kick? Learning is just one jump—no more—beyond such natal activities. Admitting that there is a kind of trigger learning linked to the earliest phase of life, there comes at least very soon another type of learning—by repetition, which however varies greatly in context. Practice makes perfect in a great diversity of circumstances. The essential here is that the response to a stimulus "alters when it alteration finds," as a result of previous experience with the stimulus. But we must go warily here. The bed of a river is the result of such varied responses to floodwaters; hillside and valley each year alter their responses to the torrent as a consequence of previous responses. Do earth and water between them "learn" adaptation to gravity? Has an old billiard table with nice easy pockets "learnt" adaptation to the balls? Is a car engine that is being run-in "learning" to do better by repetition? Have your favourite slippers "learnt" the form of your feet? Such changes are neither unusual nor particularly complex. To ennoble so common a process with the title of learning would seem a degradation of our most prized attribute. We should gain little knowledge and no credit by comparing the pinnacle of human mentality with an old shoe.

Yet we must acknowledge that, as we consider more elaborate systems, the elementary simplicities of breaking-in and wearing-out seem to merge imperceptibly into the mysteries of growing-up and growing-old. In the latest computing machines there are automatic processes which ensure that quick successful methods of problem-solving are preserved, and lengthy or inaccurate ones dismissed. Perhaps such mechanisms are only dynamic examples of running-in, but they do contain the essence of learning by repetition—with each relay or failure the mechanism alters so as to reduce the tendency to hesitate or fail in that way again.

The first failure is the beginning of the first lesson: learning begins with failure. This is a hard saying for some people, though the authors of Genesis understood it and Christian doctrine holds the Fall to have been the beginning of Salvation. Learning presupposes failure. It is only when there is a tendency to do something and the creature fails—fails any number of times—before it is successful, that learning can be of any advantage.

This is what is sometimes called trial-and-error learning, and one may be tempted here to trot out the old refrain, "Try, try, try again!"—and seek to define all learning in such terms. But does an animal try? When you see a foal getting to its feet for the first time, do you say it is struggling to its feet? Or do you say it is trying to stand up? Is that its purpose, its aim? "Oh yet we trust . . . That nothing walks with aimless feet . . ." Some purpose, some aim, is implied in the word "try." But the foal has no preconceived ideas about standing up, no image in its brain to give its struggles purpose, no ambition to stand on its own feet like the other fellows. It has a tendency to do so, a tendency to use the feet it was born with in the way provided by its inborn reflexes. It will stumble in its walk and play about until all its motor reflexes are in working order; only the successful ways of running, galloping, jumping, will be remembered; and then, all being well, it will never stumble again. But we cannot fairly accept the phrase "trial-and-error" while unsure about what is meant by try. Do your shoes try to fit? It is only confusing to inject purpose casually.

Learning by repetition, by rote, obviously plays a great part in evolution: the animal which cannot learn by repeated experience—how to get its food, how to avoid or overcome repeated dangers—does not survive. We must peer a long way back into the past for any glimpse of its origin, back to what looks like the first necessity for learning. The first life-like but immortal crystal-jelly floating in the primeval broth —not plant or animal, the virus-animalcule, our first father, now our foe—accommodated itself in an uncertain fashion to the release of pressure in the cooling mould of life. As many giant crystals do, it may well have taken many forms; and that it bred many fellows we are here to testify. The liquid crystals that joined with harder crusts of plastic skin were cells. But in these eons of turmoil, where and when can we discern learning? Adaptation or accommodation to change and stress appears with the simplest structures; in living crystals it is first seen as reproduction rather than as reflection, the bud of evolution rather than the dawn of learning.

No moment or species where learning starts can be identified; there was truly a slow dawn, a twilight epoch. As the numbers of living things increased, as species and genera multiplied, as the conditions of life became more varied, its hazards more intricate, so the inherited equipment of any animal becomes less adequate for survival, and adaptation by generation less effective. The advantage began to pass from the great breeders—short-lived, prodigal in sperm and egg, intractable creatures not of habit but of heredity—to the layers of single eggs, the long-lived docile wary beasts of the great migration. Young Diplodocus must have spent many long tropical days of playful falling and flopping in the warm Triassic ooze, learning to manage his tons of flesh and his unwieldy tail. And how many crashes persistent Pterodactyl youth must have had before winning their wings! Possibly one reason why the great Amphibia lost their empery to these playful reptiles was because the young mastodonsaurians neglected land sports, and the fate of a race was decided on the playing-fields of the Permian.

A very simple "trial-and-error" reflex was built into one of the toys described in the preceding chapter. The failures or checks of *M. labyrinthea*, limited though they are, together with their ultimate obliteration and the summation of successes, give its performance a sequence typical of learning. *M. speculatrix*, on the contrary, although in general behaviour more like a living creature, does not *learn* to follow a light or to find its way into a hutch. It is made like that. It does not learn at all. It has no mechanism for the summation of experience, no memory beyond the extremely brief and elementary one incidental to its evasion or circumvention of obstacles. Its ways are random, and that is why it seems so much more like a living creature than any other toy. What is more random than the ways of a kitten? Even in its play it is unpredictable; and in its play it is learning all the finer arts of feline chase and courtship. Whereas *speculatrix*, though he scan the scene unceasingly and seem to be looking for a bit of fun, never finds it. If he did, he too would learn, and he would have a much better time than *labyrinthea* has on its rails. But just because he is so free, if he is to learn to find the most direct way out of the room he would need a Learning Box as big as the room.

Some of our own learning, even after maturity, is of the practice or repetitive kind. Learning your way home from the station is a typical experience. But there are more varieties of the type than most psychological theorists recognise. Their favourite example of a boy learning to ride a bicycle, for instance, is an experience physiologically very different from the simple case of a foal learning to walk. Here it is not a tendency that is becoming effective, but a purpose that is being achieved. The initiative is different and the mechanisms of the whole brain are engaged; their concern with the matter subsides only after the higher centres discover that conditioning has been effected below. It would be difficult to take an EEG record of the wobbly boy, but more stable examples of a similar process of learning show very complex rhythmic effects in a brain so occupied.

Repetition of any kind traces each time more plainly a pattern in the brain, and in so doing affects other patterns throughout the organ. If we knew what happens to the brain when repetition does *not* mean learning, what happens in the brains of workers bound to a single repetitive operation, we might be reconciled to Wiener's notion of replacing them entirely by machines that cannot go wrong.

In this discussion of learning by rote it has not been necessary to introduce the notion of meaning. We have not had to ask what the foot means to the slipper, the torrent to the river-bed, or standing to the foal. We should not think of asking a parrot what it meant, even if we heard it repeat a whole column of the multiplication tables. But we might ask a child, and the answer would show whether the learning had been repetitive or analytic. When children are allowed to learn for themselves the "meaning" of multiplication, they are learning by association of one operation with another. Meaning means association. They are also learning to learn by association, and as all higher learning depends on association of one kind or another, they are likely to go further than children taught by mental drill.

Association comes early and easily to the human baby. It

is difficult to tell when it begins. It is manifest when the child
with no language but a cry begins to connect a certain sound
with a certain object. Learning of this type also depends on
repetition; but here we have the repetition of two outside
events which in certain circumstances have significance, a
special meaning. Here again there seems to be failure of a
sort before success, though it is rather uncertainty than
failure. The baby does not learn that the sound "ball" means
that familiar object the first time it is heard. The repeated
single event in learning by rote must happen regularly and
frequently, each repetition coming within what might be
called the effective interval. So also must the two events for
association, with a minimum regularity and frequency. But
here the likeness ends—the essential quality of the repetition
of the two events for association is that they should occur
each time more or less together. The sound and the object
will not be effectively associated—the word will not come to
mean the object—if they only happen together occasionally,
by chance, or so far apart in time or space that the connection between them is obscure.

Learning by association is no novelty—the notion is as old
as thought—but the ways in which it happens were misunder-
stood and misinterpreted until Pavlov undertook to measure
them. Pavlov has been neglected, travestied, reviled and
deified, but the value of his observation and ideas is undimin-
ished, provided his work—or rather the many works of his
school—be considered as a whole in the terms of the original
context. Pavlov himself was, within certain limits of his period, insatiably curious, wide-eyed, broad-minded. In his own
words, he was passionately devoted to the study of learning
as a measurable phenomenon and, perhaps most important

of all, alert to the possibility of the unexpected. In measuring the mechanics of digestion, he noticed the influence of learning; investigating the manner of learning, he came to appreciate the importance of personality. At the time when our conditioned-reflex laboratory was in operation in Cambridge the most startling of the Leningrad publications were on this subject—the identification of personality differences between the animal subjects and the elaboration of Pavlov's theory of types.

The Russian scientific workers realised at a very early stage that all animals were not equal. Today they are constrained to say that all are born equal though, as George Orwell put it, some are more equal than others. In the years of the Revolution, when floods and famine threatened both animals and experimenters, as, singlehanded, he rescued the sodden and starving menagerie, Pavlov noticed that some of his beloved beasts were impassive in the face of catastrophe, others violently agitated, others again depressed. When work began again the animals' experiences were reflected in their conditioned reflex measurements, each according to his kind. Furthermore, the extent to which an animal could be aided by sedative or stimulant drugs, Pavlov found, also depended on their type. The ones he called "weak," who were upset by disappointment and crazed by disaster, were wonderfully soothed and strengthened by minute doses of sedatives— bromides were his only stock. The "stronger" types, unmoved or excited under stress, were unaffected by these tiny quantities of sedative and merely became drowsy with larger ones. From these gleanings of disaster grew a complete theory of types, a theory so definitive as to seem dogmatic, so liberal as to be ignored in the authorised version and positively denied in the revised version of Pavlov now embodied in Bolshevik doctrine.

By unfortunate coincidence Pavlov chose as names for the four commonest types of dog, the ancient terms of Hippocrates: Sanguine, Choleric, Phlegmatic and Melancholic. The first three of these are varieties of strength, the fourth the embodiment of weakness; their recognition depends upon accurate measurement of three characteristics of behaviour, or, as we should now say, three parameters: strength, balance and versatility. All these terms present great difficulties in translation; Pavlov knew this and chose the hippocratic terms because he felt they might be more intelligible than slavonic ones, and he disliked neologisms. Intelligible they were indeed, but in unfortunate association with an obsolete and sterile system of medicine. We can see now that it would have been far more helpful to coin new words, defined in terms of the three measureable variables, and to admit at the outset the existence of other, rarer types.

With three characteristics such as the orthodox Pavlovians recognised, all of which may be present in greater or lesser degree, eight "types" are possible. The Russian workers admitted this, and in 1937 claimed to have found examples of them; the rarer four seemed to represent natural neurotics. The four types named so unfortunately by Pavlov are defined thus: Sanguine = strong, balanced, versatile. Choleric = strong, unbalanced, versatile. Phlegmatic = strong, balanced, unversatile. Melancholic = weak, unbalanced, versatile. This sounded ingenuous, arbitrary and improbable as a physiological classification of types of learning behaviour, but in fact the demonstrations which were given to us in Cambridge were convincing. Rosenthal, who initiated us to these mys-

teries, was able to predict to a few drops of saliva the behaviour of a fresh dog in the conditioned-reflex chamber,
merely by playing with it for a few minutes in a rather special
way. For example, he would snap his fingers till the dog came
to him, then suddenly would yell at the top of his voice; if
the dog paid *more* attention to the louder noise it was strong,
if it reversed its behaviour and ran at the yell it was "weak."
Similar simple rule-of-thumb tests sufficed to establish the
characteristic point on the other scales, hence to predict the
type and the conditioned-reflex patterns.

More impressive was the report of evidence which the
Russians had found, that these types were genetically determined; but the Mendelian characters responsible for them
were just being isolated when Mendelism and typology both
were jettisoned at the command of the Russian Academy.

This summary of some little-known aspects of Pavlovian
physiology should demonstrate that the basic observations
and ideas were far more acute and flexible than those of synthetic behaviourism. Analysis by types is still a reasonable
method; problems of personality and individual characteristics can be approached quite safely in this way, for there are
familiar statistical techniques to warn of approaching dissolu-
tion from types to tendencies.

Described in such laconic terms, the study of conditioned
reflexes would seem to merit the quip which Bernard Shaw
put into the mouth of the Black Girl: "Is it worth losing your
own soul and damning everybody else's to find out something
about a dog's spittle?" The matter goes deeper than this. First
there is the enormous merit of measurement as opposed to
casual observation. It was partly to ensure accurate measurement that Pavlov chose, out of all others, the salivary reflex

for his first studies. Measurement of such processes provides the experimenter with an estimate of *meaning*—a talisman sought diligently by students of semantics and others in the field of the humanities. Second, the process of conditioning extends to all levels and embraces all departments of bodily function. Once he had established the general principles of conditioning, Pavlov and his colleagues turned their attention to many other examples of association. His lead was followed in other laboratories, chiefly in America, where Gantt, Liddell, Masserman and others have studied the formation of conditioned reflexes in many species of animal and in man. It must be admitted that not all the followers of Pavlov have been able to maintain the high standards of his technique. It is only too easy to arrange a conditioning experiment—or to call an experiment by that name—and neglect the first duty of measurement. This is to fail to put the stamp on the letter, so to speak; the information reaches its destination, but the reader must pay—in confusion and enhanced scepticism, which only precision can allay.

However varied the quality of such work, one fact of prime importance emerges; there is no physiological limit to the power of association. For an animal such as man, anything that affects the nervous system can come to "mean" anything else. This association process in ourselves is so familiar that the industry of the later Pavlovians may well seem obsessional over-emphasis of the obvious. But our awareness is only a fractional measure of our associative powers; by far the greater part of this learning is below the water line, submerged in what we may still call the unconscious. The salivary reflex that Pavlov studied is beyond voluntary control; we cannot dribble at will, or stop our mouths watering when we see a lemon sucked. By studying such a reflex in animals Pavlov ensured that his measurements would not be interfered with by what the animal might "want" to do. He did not, and we should not, assume carelessly that all animals act in human fashion with conscious motives. We may find some proof that will legitimise the use of these words for other than human behaviour, but to introduce them unnecessarily begs questions that cannot yet be answered.

Involuntary functions of the body are, by definition, difficult to study in oneself, and most people are content to leave them alone. That we can safely do so is the end result of the evolutionary process already outlined, the selection of types and conditions in which the brain is liberated, in part at least, from domestic chores. We do not have to manage our breathing or our liver or the circulation of our blood. But, so intricate and mutually dependent are all the functions of the body, we can *learn* to control them if we really desire to do so. What is more, and usually worse, any of these involutary functions can learn to control "us"—that is to preoccupy, dominate, even finally to corrupt the integrity and placidity of our conscious personality.

The first set of these possibilities—conscious control of usually involuntary functions—is the basis of grotesque cults, commoner in Africa and Asia than in the Western world, in which long years are spent in practising a system of conditioned reflexes whereby the pulse rate, breathing, digestion, sexual function, metabolism, kidney activity and the like are brought under conscious control. Strange feats are performed by adepts—the heartbeat can be slowed and attenuated almost to vanishing point, the body temperature can be reduced, breathing becomes imperceptible; the whole organism is reduced to the state of a hibernating animal and can similarly be buried alive for days. The reflexes which normally prevent intense pain can be diverted so that nails can be driven through the limbs; the sympathetic nervous system can be induced to local activity, producing pallor or flushing at will; and in the same way bleeding can be prevented. The pupils can be dilated or constricted so that visual impressions become inordinately brilliant and out of focus or dim and sharp.

Such feats are commonplace to the Yogi and in some countries they are an accepted source of income. The holy man on his bed of nails is a figure of fun to Western eyes but he earns his rice by conditioned reflexes, just as surely as a performing seal earns his fish by playing "God Save the Queen," or a medical student his degree by knowing "what is *meant* by anaphylaxis." If one does not possess workshops and warehouses a new conditioned reflex is a better investment than a better mouse-trap. One who unconsciously exploits these possibilities we call original; in the East the conscious practitioner is the fakir or guru. Where economic conditions are highly developed the inventor creates a material way of life, where they are backward the fakir invents a state of mind. Both can be exploited and both are dangerous to the less sophisticated members of the community.

Many of the achievements of Yoga have been studied and imitated in the laboratory as measurable conditioned reflexes. Physiologists have spent weeks learning to make their hair stand on end or their pancreas secrete more insulin, and so forth.

The other side of the medal, the domination of the personality by involuntary functions, is reflected in the clinic where psychosomatic disorders plague the physician and delight the psychologist. It has been claimed that in every ordinary illness, even in the case of accidents and "bad luck," there is a powerful and subtle psychic element. Some disorders—ulceration of the digestive system and heart failure, for example—are certainly connected with the state of mind and influence it in turn. A person who is worried—perhaps with good reason—about his job, may start to feel pain in his stomach; the pain increases his worry and so may complete what seems a purely vicious circle. Before stigmatising so influential and widespread a power as vicious, however, we should recall that the hastened and more dramatic failure of the individual does in some cases bring about an earlier, more insistent appeal for help or mercy. On the other hand many such miseries may be merely atavistic. In our savage past, food was the first and most frequent anxiety. A high blood-pressure and rapid digestion are conceivably advantageous in a short hungry life, though they prejudice survival in creatures whose troubles are more symbolic and protracted. In the lean years foretold by the prophets these ancient associations may once more determine our vitality.

Whether we encounter these phenomena as tedious laboratory experiments with dribbling dogs, as traditionally secret physiological conjuring tricks, as an association of ideas, or as a clinical malady, the same principles appear, the principles which the Pavlov school have so abundantly established by repeated measurement. They can be enunciated quite simply in the terms of the original Pavlov experiment. Indeed this is the only trustworthy way of considering the first principles, because if we are dealing with a response or behaviour pattern which is actually itself a conditioned response, we shall find confusing relations which seem anomalous at first sight. In the simplest Pavlov experiment, the first measurement is the amount of saliva produced by a hungry dog when it is given a fixed quantity of food. This is a measure of the reflex response, involuntary, unconditioned and remarkably constant. When this quantity has been established by repeated observation, the dog is provided with a new sensation, something which either it has never had before or has experienced only in situations where it had no special significance. This may be any stimulus—light, sound, touch, smell, taste—anything that affects the dog's nervous system, either through its senses or its blood or directly by stimulation of the brain. For this stimulus to become associated with the food-saliva reflex it must be given either with the food or before the food; if the association is to be built up quickly, food must never be given without the neutral stimulus, the neutral stimulus must never happen without being followed by food, and the neutral stimulus must not be given *after* the food.

The measurement of conditioned reflexes was an important step in our understanding of how the higher centres work together and negotiate with the rest of the body. But the knowledge which these measurements by themselves can give is limited—in much the same way as if Harvey had discovered the circulation of the blood without asking how it circulates, how it is made to circulate. Pavlov's life was devoted to asking the question, how much? He jumped at suggestions for more accurate means of measurement when he was in England, but disavowed any interest in the question of how the measured events occur. And unfortunately his Russian disciples seem content to follow him to the grave of research, repeating his experiments without ever, so far as is known, asking themselves how it happens—how the animal decides that a signal is memorable, meaningful, worth responding to—a decision which must depend in some way on the relation between the new neutral signal, the bell, and the specific signal, the food.

How, then, does the brain decide that a particular pair of signals are significantly associated, that the pattern-relation of these two signals, among all those it is receiving, has some special meaning?

This question recurred years later, like an echo of Pavlov's visit, when the work stemming from Berger's discovery of brain rhythms reached a technical crisis—namely, with the construction of the toposcope, described in the second chapter. The extensive view of the physical phenomena of mental events, seen simultaneously in 22 areas of the brain, which the new apparatus provided, had taken us far beyond the fragmentary bits of mirror to be found in the scribbling of EEG pens. Many new lines of research were suggested by the sight of its "trains of travelling sparks hurrying hither and thither." The mechanism of learning was one of them.

Before developing this theme further, we may briefly recapitulate some of the EEG features which can easily be shown to be related to mental states—though in a negative fashion—and so to learning. The alpha rhythms are certainly associated with the forms of ideas, with the nature of the images set up by the thinking brain. Recent experimental and statistical studies have confirmed earlier claims that visual imagination and alpha rhythms are mutually exclusive. There is a similar correlation between theta rhythms and pleasure. In many normal adults and most children and psychopaths, the cessation of a pleasant sensation is associated with theta activity. In many cases this is in a highly stereotyped form—there is nothing so monotonous as frustration. The faster components, sometimes called beta rhythms, are well known to be common in tension, whether acute or in the form of chronic anxiety states. In considering the importance of scanning and the useful results of flicker technique, it has been suggested that the more regular familiar rhythms represent a process of hunting for information. They are the repeated asking of a question; their rhythmicity is a sign of the perpetual quest, their arrest a mark of its ending. It is in the periods when rhythmic activity is minimal that we should expect the closest correspondence with mental states—and so indeed it turns out.

The means for studying these basic electrical phenomena of brain activity, of learning, are unfortunately intricate, expensive and hard to explain. We may regret this but we cannot avoid it. If method is to match the material of these problems, we must be resigned to complexity. But the mere possession of elaborate instruments is not in itself adequate to raise the blockade on information; they must be allowed to work freely in a wide variety of conditions, and the statistical validity of the results must not be impaired by arbitrary selection of subjects or experimental arrangements. Where so little is known, any attempt at selection or control is more likely to be wrong than right. A further condition of success is that experiments should be unhurried. The term statistical in the foregoing sentence is of special and urgent importance, for the apparently irregular brain activities can only fall into patterns as a statistical effect.

It is worth considering just what this means. If we take, say, a six-channel record so as to display the noise level of the amplifiers, all six channels will be carrying plentiful signals and no two will carry the same signal. At times there will be a brief appearance of rhythmicity. If now we pass such signals through a frequency analyser, the resulting spectra from all channels will show activity about equally in all bands, and successive analyses will be similar. If we analyse over a very long epoch the spectra will be the same; if we average the spectra over a number of successive short epochs and compare the average with the individual spectra, there will be little or no difference. The same applies to phases of signals in the various channels. We are dealing with a statistically homogeneous process—all samples give the same result if they are large enough. Our truly random signals are statistically homogeneous in time, frequency, phase, and space.

Now if electrodes are applied to the head of a waking, alert, normal person, we may obtain, at a higher level of amplitude, a set of records which look superficially very like the record of "noise." (See Figure 8.) But when we examine these records with a frequency analyser or a toposcope we do not get the same monotonous results. Certain frequency bands are more frequently occupied than others; the momentary activity in several channels has a consistent phase relation; an average over a long period does not look like any of the samples from which it was derived. In other words, we have a pattern which we can distinguish from noise only by comparing the statistics of its distribution in frequency, phase and space with the results we should expect to get from a truly random "noise."

These concepts are not easy to grasp when unfamiliar. An example from linguistics may clarify. The letters in this sentence could be written down in any order: ehtr edros rette lynan inisi htnwo decne tnesn ettir wdlu oceb. If sufficiently disturbed, the letters lose their meaning; but in the rearrangement given, the meaning can be deciphered quite quickly. A count of letter frequencies would suggest that it was not a truly random sample of letters—there are nine e's, seven t's, and so on, whereas in a random set of fifty-two letters any letter of the alphabet would only be expected to occur twice. And the longer the message, the easier it would be to decipher it.

If the EEG be considered as a cryptogram, the nature of the problem is easier to understand. The rhythmic components are the gaps and punctuation marks, common but quite meaningless in themselves; in fact they could be omitted from the complete message without much loss. In order to stand a reasonable chance of deciphering the EEG signals, we must make sure there are enough of them and that the record is reasonably complete; that is, the record must be long enough and taken with as many channels as we can afford. Then, in order to decide whether the signal is significant as opposed to noise, the quickest process is to perform a series of frequency analyses; this will not tell us what the meaning is but it will give us a good idea as to whether there is meaning or not.

If the frequency analysis is encouraging the next process is to search for substitutions and transpositions—that is, to analyse in terms of phase and space distribution. A further and quite invaluable stratagem, flicker, has already been discussed. In the cryptogram analogy this may be compared with forcing the enemy to send a certain message. Obviously, if this can be done, then the deciphering of other messages is likely to be easier since the code will be known. The response of the brain to such stimulation as we have described may be anything from a simple elementary discharge at the stimulus frequency, to an elaborate transformation of the stimulus rhythm into a widespread fabric of intricate oscillations.

With the toposcope it is possible at the same time to follow the impression made on the subject by these procedures. Thus the experimenter has three sets of data to correlate; the original stimulus as a physical event, the cipher derived from this by the brain, and the mental state evoked in the consciousness of the subject. In these observations there is an encouraging correspondence between the extent and complexity of the evoked responses and the subjective sensations. As already noted, the more vivid and bizarre the experience of the subject, the further from the visual areas are the evoked responses and the more peculiar their form and geometry. This relationship was obvious even in ordinary EEG records, but with the toposcope display the correspondence is more striking and the details are more sharply outlined.

The next feature is that both the evoked activity and its spontaneous background are intricate in the literal sense; that is, what looks like a single wave in a conventional record is nearly always made up of several interwoven components. The electrical structure is fabric-like, with the warp and woof usually clearly distinguishable. It is hard to recognise activity in more than two dimensions, but almost certainly there is a third, at least.

What, then, is the function of these elaborate electrical structures in the economy of brain physiology? What process can require so laborious a transformation? To what end does the brain construct these transient fabrics of hard-earned energy?

To answer this question we may ask another, easier to answer but often neglected. What *specific* property has the brain that other systems have not? Is the "great ravelled knot" of the brain different in kind, or only in size, from its stalk, the spinal cord?

The answer is clear and opens a wide door to truth. The brain can learn—NO other structure can. So rare and so precious is this learning, so delicate and so elegant is the electric weaving we have seen, that to associate the two is more than tempting—it seems a marriage of necessity.

CHAPTER 7

The Seven Steps from Chance to Meaning

The world is either the effect of contrivance or chance; if the latter, it is a world for all that—that is to say, it is a regular and beautiful structure.

Marcus Aurelius

AFTER GLANCING at various types of learning in the last chapter and considering some of them physiologically, two separate paths of research were found to converge in a significant manner. The problem of what is happening in the brain when a signal or an event acquires meaning through association with some other signal or event was illuminated by the conjunction. Of the two paths, far apart in source, one had started with Berger's discovery of electrical rhythms in the brain, the other with Pavlov's discovery that any bodily function can be made the basis of a conditioned reflex and that one conditioned reflex can be built on another. Both lines of research have yielded invaluable information, and each will continue to do so in its own way. It is their convergence, however, which today seems crucial in the evolution of brain physiology, because it brings together, in a natural union, a mechanism of unknown function and a mental activity as yet unaccounted for, both the mechanism and the activity being uniquely and supremely attributes of the living brain.

At this point it might be thought that the course of future enquiry would be plain—that, two main paths having converged, there would now be only one to travel. It would be convenient if scientific exploration could be conducted in this manner, like tourists converging on Calais with the white cliffs of home in sight of all. But scientific travel is not a homecoming; we arrive only to depart; and a rare point of convergence like this one has many more departure than arrival platforms.

Before going further, what has to be considered above all is the condition of the vehicles in which we have arrived and the character of the loads they carry. On the one hand there is a great store of EEG equipment which is constantly getting new information in spurts and flashes, but such that we can never be sure just how the pieces fit together until the validity of each piece is confirmed; on the other hand, a vast load of valuable data, verified and classified but unused. Should these travel together on one road, or separately on their own lines? Can we, for instance, get EEG records that show the conditioning of a reflex, and where would this take us?

Some work of this kind has been reported from Russia and France. Livanov and Poliakov, using normal rabbits, obtained EEG records in which they found effects of the conditioning of a reflex.

Any record taken while an animal is going through any such experience will certainly show not one but many strange patterns and alterations of the rhythms. And this applies *a fortiori* to records of human subjects. The difficulty is precisely in the very multitude of effects which are observed. Among them no doubt are the specific results of the particular experience. But by what process of analysis can they be disentangled from the rest? And would it, in any case, answer our question?

Electroencephalography got off to a bad start because we were mistaken about the apparent simplicity of the alpha rhythm; and that, mainly, was because apparatus of excessive simplicity was used. The same dangers are present whenever the complexities of an experiment exceed the scope of the experimenter's equipment. It is not quite clear from their reports whether the Russian and French workers were fully aware of this. With the toposcope, an instrument more appropriate to the intricacies of such observations, many effects of all kinds of stimuli or signals have been watched, spreading from area to area of the brain, but in such profusion that it must be some time, as we have said, before these cryptographic messages can even be classified for decoding. Travel on that road must continue slowly and methodically.

Meanwhile there was the vast load of complete, verified and classified messages about conditioned reflexes only waiting to be decoded. While convinced that more knowledge about the rhythmic mechanisms of the brain will be necessary for a proper understanding of the whole process of learning, we felt we should be well on the way if we could ascertain what operations the brain must go through to arrive at some associative meaning of a signal or other event. What is changing in the baby's brain while it learns the meaning of the word "ball"? The question which Pavlov was reluctant even to consider, which also, for lack of any information to the contrary, we must regard as being either neglected or taboo among his Russian disciples today, is this: What happens inside the brain when a reflex is being conditioned? How did the rabbit get into the hat?

Once again, as when faced by other problems of function, guidance was sought by attempting to construct a working model. Having already in *M. speculatrix* a model with mobility governed by a reflex responsive to light, it was a simple addition sum to provide a second reflex circuit to be made responsive to sound. The further addition of a conditioning connection between the two reflexes should then make it possible, for instance, to train the creature to come into the room at the sound of a whistle, after teaching it that sound "meant" light. This turned out to be no operation in simple addition, however, but more like practice.

> Multiplication is vexation,
> Division's just as bad.
> The rule of three, it puzzles me,
> And practice drives me mad.

So the first trials made with the simple addition of the electric circuits indicated were disappointing. It took some time to appreciate the number and complexity of the operations involved in the provision of a conditioning connection between two circuits. More information was sought. But in all the mass of empirical data available there was no hint of the character or number of operations needed. Obviously one of the principal requirements for a process of associative learning must be a complex mechanism of memory, capable of sorting the traces of the two series of events to be related. Mechanical "memories" are familiar devices. This one would have to be able to do more than store information, however; there would have to be recognition and registration of a coincidence between two events which was greater than would be expected by chance—which might be a matter of statistics.

But with no other light than this, it was difficult to envisage what the core of the problem was.

An entirely new sort of approach seemed to clarify the proposition. The transmission engineer who has to design a telephone system must first discover how signals can best be distinguished from background noise. For comparing the various devices at his disposal he has a useful trick. He gets a lot of information about a transmission unit simply by comparing the signals that go in with the signals that come out. He often calls the unknown unit a Black Box and undertakes to determine its performance by measurements at its input and output terminals, without looking inside.

Our problem may be simplified then if we begin by considering the brain centres as a generalised transmission system into which signals go and from which other signals come out. This method advantageously permits the use of powerful statistical means for ascertaining the simplest way of distinguishing important pairs of signals from random or meaningless ones. We begin, as already noted, with a mass of dependable information about the relations between effective pairs of signals that go into the Black Box and the responsive signals that come out of it. Let us consider briefly what are the exterior conditions in which we know that two signals come to have a significant effect, and how this knowledge can help us to get some idea of what must be happening inside the Black Box.

Study of learning in animals and man has shown that, whatever the final result of an association, some process of selection must take place before any outward sign of learning appears. In conditioned-reflex experiments the neutral stimulus—the bell, for example—and the specific stimulus, the food, must be given together a number of times before the new conditioned response is established. For the first few trials, the animal may seem "interested" by the new experience, but no saliva appears. If the bell and food occur simultaneously and regularly, after ten or twenty trials the flow of saliva begins when the bell is sounded alone. The size of the conditioned response is not at once as great as for the food and in some animals is never quite as great. These measurements provided two of the characters that vary very much between individuals and help to decide the "type"—namely, the length of time taken to establish the new response, and the size of it when established, as compared with the unconditional reflex response.

From such observations we can be sure that in the Black Box, as well as the *constructive* processes of memory and association, there must be a preliminary *selective* process, an operation which can determine, in effect, whether a new signal is worth bothering about. Every living creature is constantly bombarded with signals of all kinds. The brain of a human being, however simple and uneventful his surroundings, is receiving every second of his waking life several hundred sensory signals from the outside world and from the rest of the body. Any or all of these signals may have some important meaning in relation to any or all of the others. Yet somehow, for anyone to learn the simplest things, those signals which occur regularly in association with others of known or basic importance, must be sifted out from all the others from the background noise of life. In ourselves this process is often conscious, but it need not be so. It is a common experience to see a strange word or name, one thinks for the first time, and then to come across it shortly after again and again.

Nearly always in such cases it can be shown that, in fact, the "strange" word has been seen often before but not in an important or interesting context; when it appears in a more significant association, it is remembered and learned as though it were fresh and meaningful. In a previous chapter we discussed the significance of pattern—here we are beginning to discern, as it were, the pattern of significance.

In laboratory conditioning experiments, this selective operation could not be studied very carefully because the situation was deliberately arranged in order to make the choice easy for the animal. Pavlov's early experiments were made with the animals in stimulus-proof rooms—until he realised that for many animals the *lack* of stimuli was itself a stimulus! Thereafter the animals were only mildly restrained and could hear and see random events, casual passers-by, the wind and rain or sun; and provided these background stimuli were unrelated to the test stimuli, the animals learned more quickly and maintained their response more uniformly. In human affairs, too, there is nearly always pre-selection of important conditioned stimuli. This is the value of a school. A teacher selects from all possible stimuli those which he believes to be important, adjusts the background conditions to a compromise between solitary confinement and the market-place, establishes a set of unconditioned punishments and rewards, and exposes his pupils to the association of conditioned and unconditioned stimuli as regularly and firmly as convention permits. The importance we attach to the pre-selection and segregation of school is a measure of the inefficiency or inadequacy of the selective mechanisms in our Black Box. Were these built-in selectors perfectly efficient, we could learn all we need to by just letting things happen to us.

Most people would indeed maintain that the important lessons are learnt out of school and that it is only our "artificial" life that requires so much book-learning. If we had only ourselves to consider this would probably be true, but in any society larger or more intricate than a family, it is necessary to learn not merely the facts of life but how these facts are viewed by other people even far away and long ago. *Necessary*, not merely desirable, because these opinions of others may be as powerful and important in deciding one's fate as the leap of a tiger or the beat of one's own heart. In the last chapter it was suggested that learning begins with failure— and that practice makes perfect a successful change; now we may surmise that teaching becomes necessary when the association of events is too uncertain for the built-in guessing-machine of the brain.

Learning by association involves guesswork. How can the Black Box guess? On what does it base its conjecture? Unaided by a teacher, exposed to all the chance and change of an indifferent or hostile world, what feature does the Black Box look for among all others?

Here again the details of the conditioning procedure give us an essential clue. If an association is to seem significant, the neutral signal (say, the bell) must always be either simultaneous with or precede the specific signal (the food). This is only like saying that if the dinner bell is to be any use it must be rung before and not after dinner. But the time relation between the two signals—bell and dinner—is critical. If the neutral signal appears a long time before the specific signal—that is, if the bell is rung, say, an hour before the food comes on the table—the sequence of the two signals will have to be repeated much more often to give the former significance than if they are simultaneous or there is only a short interval between them. It might take you a number of days to realise that a bell had anything to do with dinner if it only rang an hour before the event. The next thing to note is that if, after the association has been established, the coincidence of signals never happens again, the memory of the association will fade away—that is, if dinner is always late you will stop paying any attention to the bell. And the memory of the association will fade still more quickly if the bell rings and there is no dinner after all. We can recognise such an experience in ourselves as a feeling of disappointment or disillusionment.

This relation between bell and dinner is not a simple reversible one—"bell means dinner" is not the same as "dinner means bell." We cannot apply our simple arithmetic to this relation. Thus, if, after a conditioned response has been established, the food appears without a bell, the conditioned response will diminish, but at a different rate, and it can be re-established more easily. We can feel this in ourselves, not as disappointment but as indifference due to satiety. If we are well fed, anyway, the dinner bell does not have any regular or insistent appeal; but if we miss a meal the signal will become important, however content we were to ignore it before.

For most people awareness of these relations comes most vividly from experience of training animals or watching their performances. When an animal has done its trick it is immediately given a reward—usually food; when it comes on the stage it is always hungry. The food must not be given before the act, nor too long after it. But we all have seen some animals go through trick routines without apparent rewards—the circus horse is not ostensibly fed with oats or sugar, nor the lion with hunks of meat. There are two ways in which this can be done;

the first and more difficult is to establish a second conditioned response alongside the original one—that is, to teach the animal that a pat or an encouraging word means food and that doing the trick means a pat—the proposition is then complete; if trick, then pat, then food. The second way of dispensing with immediate reward is to replace "if" with "unless"; that is, to punish failure instead of rewarding success. The punishment need not be very severe. The claim of animal trainers and schoolmasters that "it's all done by kindness" may often be true; but it would be a hard task to break in a wild horse with lumps of sugar.

Learning to associate two events when the later one of the two is painful, is sometimes called a "defensive" reflex as opposed to an appetitive one when the second satisfies an appetite. The distinction is not very clear since one might think of an association such as "if bell then food" as equivalent to "unless bell—hunger"; and starvation may seem as stern a discipline as the whip. Most of us have been sent to bed without our supper, and some of us have been whipped; which was the more memorable, which did us more good, and which was the more cruel? Popular opinion is divided here—almost evenly divided.

Human experiments are obviously impossible to devise, but animal ones have given fairly clear answers. If a neutral stimulus is given before a painful one and the pain is more than just discomfort, the association can be established by a single experience and need never be reinforced. This implies that the avoidance of pain is, by itself, a sufficient unconditioned stimulus, needing only an occasional formal reminder. A crack of the whip or the whiff of grapeshot may be more potent and more thrifty than reiterated rewards.

There is one serious disadvantage of this system—its effectiveness varies particularly widely with the type of subject, whether animal or human. Pavlov found this at an early stage, and everyday experience confirms it; there are children—and adults—who can learn quickly the hard way, from one harsh trial, and will never forget and will never resent the discipline. But there are others not less numerous and as apt to mastery, whose response to stern discipline is inverted and rebellious. Many dogs who have achieved the highest degree of trained competence in the laboratory by a scheme of carefully graded and timed rewards, break down into surly treacherous curs when punishment for failure replaces reward for success.

This innate intolerance of retribution is associated in animals with other features of what Pavlov called, with misleading disparagement, "weakness," a character typified also by an inverted response to very powerful stimuli. If some selection board of the future wishes to predict which recruits will respond best to the guardhouse and which to promotion, they might do worse than substitute a demolition charge for "Come to the Cook-house Door"—the men who can still empty their plates will tolerate stripes on the back when they do wrong; the others are likely to benefit more from having stripes on their sleeves when they do right.

Consider now in greater detail the mechanisms we should expect to find in our Black Box. It certainly must contain a good deal more than meets even the inward eye. Clearly, before an association can mature, there must be a device to sort out and select incoming signals on the basis of the order in which they occur and the regularity of their coincidence. The end result of this operation is to arrange pairs and clusters

of events in a scale of meaning; those that come in always
exactly together and never separately will be at the top, those
with random irregular associations at the bottom. There is
plainly a statistical analysis; and if this seems too academic
and pedestrian a conclusion we may add a tinge of romance—
or squalor—by noting that what we should expect to find
just behind the input terminals of our Black Box is a Book-
maker.

What we are suggesting is that the brain must, quietly, unobtrusively, incessantly, reckon the odds in favour of one
event or one set of events implying another. In the simplest,
isolated case, as in the laboratory, the odds are based on form
—on the past history of that particular twin set of events being considered, on how they have come in before. But in
more complex natural conditions the resemblance to the racecourse becomes even closer, for a creature may have a wide
choice of events and like any gambler can back a short-odds
favourite or take a longer chance on an unpopular outsider.
On the favourite—that is, on the events that are obviously
related—all backers have some money; the rewards are small
but fairly certain. On the long chances there is less competition, more risk, a longer wait, more anxiety, but the hope of
a fortune.

In the evolution of living creatures can be seen the signs
and effects of the struggle for the daily double. The animals
that are well adapted and specialised—worm or fish, mighty
lizard or mastodon, those who for some epochs were lords of
the earth—were cautious backers of the dead cert; those
whose nervous systems lay fallow during long ages, subdued
by their mighty neighbours, starfish or newt, little furry creature or heavy-headed ape, were all the time quietly develop-

ing a system to pick out dark horses; and from time to time the meek did inherit the earth.

> I returned and saw under the sun that the race is not to the swift nor the battle to the strong . . . but time and chance happeneth to them all.

Time and chance. We underestimated their part in the processes of learning before we began to suspect that the brain is a bookie and the Preacher's God a mighty gambler. Our early failures to confer the power to learn on our models were due not merely to ingenuousness but to sheer ignorance of how much dealing with time and chance is involved in learning.

Finally, in the light of such ancient wisdom, and with the more detailed knowledge of the statistical pre-selector, given by study of the learning brain as a Black Box, we were able to specify the elements of its interior mechanism and to construct a model embodying them. It was then found, in effect, that this model could learn as animals learn, forget as they forget, and even sulk and fret as animals do when things are too much for them.

The operations a model must perform can now be enumerated. First, there must be a device to differentiate the *beginning* of an unconditioned specific stimulus—the appearance of the food in the simple example—from its continuance. This is to ensure that a neutral stimulus which happens to occur *during* dinner is not taken to *mean* dinner—otherwise we should always be late for table. This is a straightforward operation and can be represented diagrammatically. In Figure 12 is a diagram of our Black Box in its simplest form, with two transmission systems, T_1 and T_2. T_1 is the reflex which oper-

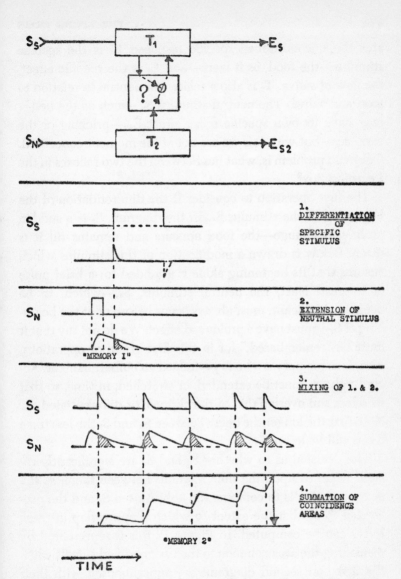

Figure 12. ". . . what lies between the two reflexes in the Learning Box?" The first four of the seven operations of learning—the selective operations.

ates the unconditioned specific response. S_s is the specific stimulus—the food, as it were—and E_s is the specific effect, the flow of saliva. T_2 is also a reflex, but neutral in relation to food and saliva. The neutral stimulus S_n—such as the bell—may have its own specific reflex effect E_{s2}—pricking of the ears, say—but with this we are not at the moment concerned. Our main problem is, what lies between the two reflexes in the Learning Box?

The first operation to consider is the differentiation of the beginning of the stimulus S_s. In the diagram, S_s is a sudden prolonged change—the food appears and remains till it is eaten. Below is drawn a modification of this stimulus which ensures that its beginning alone is attended to: a brief pulse of interest. Next, the neutral stimulus, S_n,—which, to be deemed significant, must always occur before or at the beginning of S_s—must have a prolonged effect. We do not say that it must be "remembered," for in this first group of operations, conscious memory is a luxury—indeed an embarrassment. Accordingly, S_n must be extended, or stretched, in time, so that its effect can overlap that of the clipped, or differentiated, S_s. Note that the longer the delay between S_n and S_s the less trace there will be left of S_n.

The decision as to whether S_n and S_s are significantly related depends upon the time relations between them; so the next operation is to combine the abbreviated S_s and the protracted S_n so that the extent to which they overlap in these forms can be computed. In Figure 12 this is represented by measuring the area common to the two modified signals when the first and second diagrams are superimposed with their correct time relations.

The measure of significance also depends upon the number

of times S_n and S_s have been associated; therefore the next operation must be to add up all the areas of coincidence that have been observed from time to time, so that a consolidated measure can be arrived at. Since, however, the absolute interval between experiences is important, the significance attached to each operation must be allowed to diminish slowly with time. If the experiences are separated by too long an interval their effects will not summate, since only an infinitesimal trace of the previous experience will remain. The progressive building up of the measure of significance can be represented as a graph on which such coincidence is seen as a rise in the line followed by a slow decline as its relative importance diminishes. If the extent of coincidence is great at each experience and the experiences are frequent and close together in time, the graph rises quickly to a given level. If the degree of coincidence is less or the experiences are less frequent or further apart, the graph rises more slowly and may never reach the threshold value.

The next point is to decide what this threshold should be. What *is* a "significant" degree of coincidence in a series of observations? How close together must S_n and S_s be, how often must they occur, and how long can we wait between experiences? Here both everyday experience and formal statistical analysis come to our aid. Statistics indeed arose from just this sort of problem, from the study of games of chance, from gambling. If you were playing at double or quits for instance, with a chance acquaintance, how many times would he have to toss a head before you would feel safe in accusing him of using a double-headed penny? Common sense—your own store of statistical estimates—tells you that you may often get three heads in succession quite fairly; five, quite often; ten,

—rather unusual; twenty—well, really; a hundred—never! What, *never?* Well, hardly ever. This is where the difficulty arises—you can never be absolutely certain that your suspected playmate is not just extraordinarily lucky.

This of course is the lure of gambling—there is no reason why you too should not toss twenty heads in a row. At Monte Carlo a number of roulette wheels have been spinning for many years. As well as the numbers, there are several even-chance features, red or black, odd or even, less than or more than 18. Like tossing a penny, each spin of the wheel is independent of the others; the way the ball runs one time does not influence and is not influenced by the other spins. In the whole history of such games there has never been a run of as many as 30 in the even-chance events. The Casino rules of play are based on this. If there were no minimum stake and no maximum, a gambler could wait for a run of, say, five blacks, then put one franc on red; next time, if black came up again, he would put 2 francs, then 4, and so on. In this way he could continue to double his stake and be sure that when at last his colour came up he would win the value of his first stake. If he had a run of ten adverse colours, he would have staked just over 1,000 francs in the hope of winning one—a dull game. But if everyone played it, and there were no minimum and maximum—and no double-zero—the bank would slowly but surely lose. The ratio between minimum and maximum stakes is usually arranged so that, starting at the minimum for the table, eleven or twelve doublings reach the maximum: one to a few thousand. Since the limit is generally about an average professional income for a year, the game is arranged so that by the doubling method you may win the price of a packet of cigarettes though likely in the end to be ruined. Actually the best chance of winning is to put all one can afford to lose, once, on one number—a 36-to-1 chance—and then, win or lose, go away.

How does this help to establish the odds that our Learning Box must accept as significantly in favour of a series of coincidences being more than chance? Leaving the racecourse and Casino, we may return to the laboratory to see what an animal sets as standard of significance. As already mentioned, in the simple isolated conditioned reflex situation about twenty trials are usually necessary before the new connection is established. In general, and allowing for differences between types and conditions, the number of perfect coincidences an animal requires before its brain says "perhaps" is more than ten and less than one hundred. So the mechanism which performs the first four operations of learning, the statistical filter, may be adjusted so that the threshold of significance is just attained by, say, twenty experiences in which S_n and S_s are exactly coincident and happen as rapidly as the response to S_s can permit. Thus, whatever the nature of time and chance, at least our model will be lifelike.

As a further check on this estimate we can give one assistant a key to press which delivers S_s and to another a key for S_n. The two operators, in separate rooms, are told to press their keys as and when and as often as they like. The operators, being independent of each other, provide a series of randomly related signals; the statistical filter should only very rarely build up the threshold value of significance from such an input. Again, however, we can never be quite sure that by "sheer bad luck" even these independent sources will not turn out a series of coincidences. But once more everyday experience is helpful. Every day, without any thought of risk or danger, we take chances of a million to one against violent death. In a month of ordinary travel we accept with greater reluctance a hundred thousand to one in our favour. At ten thousand to one we are conscious of the risk, but take it, none the less. At a thousand to one against disaster we feel excited or indignant, according to whether we happen to be at the wheel or on the roadside. Risks of a hundred to one we expect to be paid well for, and at ten to one we are decimated and can count on a decoration. So if our Learning Box needs twenty bull's-eyes to get an idea it must be so built and set that at the opposite extreme the odds against it reaching its threshold by chance are greater than a thousand to one. To maintain a lifelike attitude it must be adventurous, but not reckless.

But what if the neutral stimulus S_n is kept on all the time, if one of our operators decides, as his instructions allow, to keep his key pressed down all the time? This obviously makes S_n meaningless, and we must therefore arrange for S_n also to be effective only at the beginning, though protracted as well to a degree decided by the considerations we have just unraveled. It will be seen later that this essential specification of S_n has important secondary consequences when we consider how to look for the living processes themselves.

Assuming now that our Learning Box has thus built up the threshold value of coincidence, the next operation is to transfer the information—that S_n and S_s seem to be associated— to a long-term storage register. We can call this operation Activation. (See Figure 13.) Observation of animals suggests that it is often abrupt, and in ourselves it may correspond with the awareness of something having been learned, or with "getting the knack" of some problem.

The next operation, which preserves the information of

coincidence, may more legitimately be dignified with the title of Memory. In this operation, too, provision must be made for the possibility of error—the association may have been chance. So the memory must be allowed to fade quietly away if the supposed association never occurs again. It would never do to have the whole Learning Box preoccupied with what might have been.

The last operation is the combination of the memory with a fresh sample of S_n. We may say that, once the preservation system is activated, it opens a gate through which a signal may pass from S_n to have the effect E_s—and the reflex has at last been conditioned. (Figure 13.)

No wonder our simple-minded experiments were of little avail! Who would have thought the Box had so much in it? And all these stages are essential; if any of them is left out our Box will imagine vain things—or nothing.

To summarise, the first two of the seven steps from chance to meaning are strictly selective; the second two are temporarily constructive; the fifth is a trigger; the sixth a preservative; and the seventh effective. In comparing experimental animals, machines and men, one must be duly cautious of meretricious resemblance—mimicry is not explanation, and a model is not a mould. But when more than three or four properties of a system are known and the simplest possible model which could reproduce these properties has been constructed, consideration of the hypothesis, that the original contains components comparable with those in the model, is permissible.

The next step is to test this hypothesis by prediction. Having studied our Black Box and built up a model to act as we think the Black Box must act, what other properties should we expect the model to possess, from what we know of its components? The complete circuit of what for short we call CORA, the Conditioned Reflex Analogue, will be found in Appendix C. CORA grafted onto *M. speculatrix* produces a new species, *M. docilis*, the easily taught machine, which be-

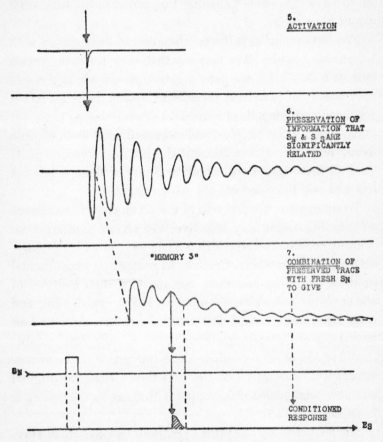

Figure 13. "What becomes significant . . . is a private image . . . an idea." The last three of the seven operations of learning—the constructive operations.

haves astonishingly like an animal. It can be taught to come to a whistle by blowing a short blast and showing it a light—light in *M. speculatrix* eliciting an unconditioned reflex of attraction. This must be done fifteen or twenty times within a few minutes before any new effect is produced. If the whistle is blown all the time it has no effect; and the light too must not be on all the time. There is no effect of course if the whistle is blown after the light is shown. If the whistle is blown half a minute before the light is shown, conditioning may take hours. Once a new response has been established it fades away, if it is not reinforced from time to time; and if we tease the creature by blowing the whistle without the light, the response will diminish more rapidly, though a latent memory may persist.

So far, so clear: we have reproduced nine or ten of the features of animal learning by arranging three vacuum tubes and sundry other components so that they perform seven operations. This is a promising analogy but, so far, no more. We could have made the whole thing from clockwork or chemical reactions; the electrical basis of the model should not in itself prejudice us in its favour. It is compact, economical and does all we expected. But does it predict anything? Does it contain other properties than those we found necessary to specify its simplest design?

In experimenting with *M. docilis* we soon tired of the sound-means-light trick and decided to teach the model that sound means trouble. This we did by connecting CORA to the obstacle-avoiding device in *M. speculatrix*, the feedback circuit which made that simple creature back and turn when its shell was touched. Its education by punishment consisted of blowing the whistle and kicking the shell a few times. The electrical signals thus generated are much larger than those produced by light, so fewer experiences were needed. After five or six beatings, whenever the whistle was blown the model backed and turned from an "imagined" obstacle.

Then came the first unexpected confirmation of the similarity hypothesis—it could have been foreseen, but was not. The activation operation in CORA consists of a brief discharge in a miniature neon glow tube—it is seen as a flash of pink light. Now when we were trying the effect of blowing the whistle after the defensive reflex had first been conditioned, we saw, without reinforcement by kicking, the flash of light which indicated that the memory circuit had been activated afresh without the specific stimulus. This had never happened of course with the sound-means-light reflex; but the circuit is a probability analyser—it could happen by chance. But every time the whistle blew and the model dodged its imaginary impediment, the pink light flashed. Then we realised that so it must turn out; the defensive reflex involves an internal feedback circuit in which signals circulate from the output to the input of the light amplifier. This means that *M. speculatrix* can pay no attention to lights when it is dodging an obstacle, but it also has the consequence that, when it has learnt that "whistle means dodge," its very dodging instantly sends a specific signal back into CORA, and the memory of the learned response is automatically reinforced. The defensive reflex does not require specific reinforcement —exactly as in the animals and ourselves. Had nothing been known about the main difference between appetitive and defensive reflexes, of the relative value of reward and punishment, it could have been predicted from the uncalculated behaviour of the model.

Another test of the fitness or goodness of fit of the model is that, if added frills extend the inventory of accessories, the lively extravagances of animal and human experience should snugly fill out the swelling fabric—the flesh should match the metal, not in Procrustean fashion but as to the nature born. This is confirmed when more degrees of freedom are added to *M. docilis*. The simplest model provided for only one contingency at a time—"sound-means-light" or "sound-means-dodge"—and we assumed no impediment to its realisation. But there is a possibility that in some other circumstances we may have to weigh novelty against custom or consider the less familiar proposition that light means sound or touch means sound.

This situation is more easily envisaged if we consider first the reflex, built-in, effect of sound. It was mentioned that the neutral stimulus, the sound of the whistle, might have a direct effect of its own. In a real animal a sudden sound would in fact be likely to evoke some reflex action; commonly an animal will "freeze" to a sudden sound. We can, therefore, without loss of superficial realism, and obeying our accepted laws of mimicry, connect the output of the sound amplifier to the relays controlling all the motors in the model, so that when the whistle is blown all the power is cut for ten seconds or so, and the model cowers motionless and silent, "plays possum" for this period, then crawls quietly away. Now, when the attempt is made to establish the conditioned response, sound-means-light, we find that we must add an *inhibitory* link from the final gate of the Learning Box so that the newly significant positive association shall not conflict with the reflex inhibition of all movement. Thus we have introduced "inhibition of inhibition."

But the model with this character is not quite so perfectly docile as the original and a new possibility of variation is presented. The threshold for "freeze" may be made higher or lower than that for the conditioned response. If it is made higher, then the acquired association will be prepotent—after training, sounds of ordinary intensity will always be attractive, though very loud sounds are likely still to evoke immobility. If the reflex threshold is lower, ordinary sounds will remain inhibitory but loud ones can become attractive.

In practice the setting of these thresholds is very critical and varies with wear and tear and the state of the model's batteries. Such models show inexplicable mood changes. At the beginning of an experiment the creature is timid but accessible, one would say, to gentle reason and firm treatment; later, as the batteries run down there is a paradoxical reversal of attitude; either the reflex or the acquired response may be lost altogether, or there may be swings from intractability to credulity. Such effects are inevitable; however carefully the circuits are designed, minute differences and changes are cumulatively amplified to generate temperaments and tempers in which we can see most clearly how variations in quantity certainly do, in such a system, become variations in quality.

When, having sampled the possibilities of a single Learning Box, another is added to provide for the reversed contingency already contemplated, we observe many strange things. We can now establish the association sound-means-light as before, or light-means-sound, or touch-means-sound. For the first association the response is "go to sound"; for the second, "freeze to light"; for the third, "freeze when touched." Now suppose we establish the "sound-means-light" response, but, while the response is being carried out, we rebuff the creature's advance to the confirming light with a kick, and repeat this from time to time. We are building up the association "touch-means-sound, so freeze." But once this conflicting contingency has been established, we find that any stimulus entering the transmission system for light will have the same result as a sound-paralysis. In trying, as it were, to sort out the implications of its dilemma, the model ends up, "sicklied o'er with the pale cast of thought," by losing all power of action. It is no longer possible to tempt it by gentle light to its feeding hutch where its batteries can be charged; nor can a sound mean light, for any attempt to reinforce this original lesson will itself induce a palsy. Of course, were the model an engine for guiding a projectile or regulating the processes of an oil refinery, this tendency to neurotic depression would be a serious fault, but as an imitation of life it is only too successful.

In laboratory experiments, conflicting incentives and penalties have been studied with industry and imagination by the orthodox Pavlovians; also by Masserman in Chicago, who, as a psychiatrist, was deeply interested in the sources of mental pain. In cats, he found, quite elaborate responses can be inculcated when food is the reward. A normal cat can learn to press a bell three times to get its food delivered, but if then a puff of air is directed at the cat's face as it darts to the food, the situation changes dramatically; its manner quickly becomes sulky and suspicious, its skill is lost with its appetite. Often it turns its indignation to the bell-push it once manipulated so deftly, and scratches and bites it. Many cats refuse to feed, grow thin and ill, and would die rather than submit to the doubt and indignity of the harmless puff of air; an insulting reproof in the pay packet has caused many a strike. Masserman was most intrigued to find that in most cats a few drinks would dissipate these fine airs. The discouraged animals soon learned to pick out the hard liquor from the soft drinks, and attacked their food trough with abandon. But unlike some humans, once they had broken down their mental barrier, they could take their liquor or leave it.

Such transformation can be imitated with great fidelity by CORA, as can the other everyday observations of psychiatrists. When a complex learning model develops an excess of depression or excitement, there are three ways of promoting recovery. After a time the conflicting memories may die away —except in obsessional states based on defensive reflexes, which as we saw earlier tend to be self-sustaining. Switching off all circuits and switching on again clears all lines and provides, as it were, a new deal for all hands. Very often it has been necessary to disconnect a circuit altogether—to simplify the whole arrangement. Psychiatrists also resort to these stratagems—sleep, shock and surgery. To some people the first seems natural and benign, the second repulsive, and the third abhorrent. Everyone knows the benison of sleep, and many have been shocked into sanity or good sense; but the notion that a mental disorder could be put right by cutting out or isolating a part of the brain was an innovation which roused as much indignation and dispute as any development in mental science. There are volumes of expert testimony from every point of view, but our simple models would indicate that, insofar as the power to learn implies the danger of breakdown, simplification by direct attack may well and truly arrest the accumulation of self-sustaining antagonism and "raze out the written troubles of the brain."

Some reference, necessarily superficial, has already been made to the relative numbers of brain cells and brain functions. If N represents the number of functions, then, to be capable of learning any sort of contingency or association between any or all of these functions, we must have $N^2 - N$ learning boxes. It was suggested that the number of elementary brain functions may be of the order of 1,000, and each of these may need about 100 cells to provide for the average range of intensities and amplitudes of sensation and performance. For these 100,000 elementary functional departments to have full scope for learning about one another's experience, starting with the "anything may mean anything" assumption, the necessary learning links just about account for the number of cells we usually have in our heads.

If such a system were really in the least like *M. docilis* multiplied by a factor of a few thousand million, it could not work long or well without several precautionary and safety devices. There is the danger of everything seeming to imply everything else and nothing being what it seemed. How such risks may in fact be guarded against without reducing too drastically the power of association will be discussed later. But considering our models, we can find some encouragement in the figures of "nervous breakdown" in ourselves. In most countries, one in ten of the population needs or seeks expert attention for mental trouble at some time, and a much higher proportion takes some mild treatment or medicine—a rest cure, sedative, stimulant. So, comparing metal with flesh, it is not necessary to design for a factor of safety; as in the case of a military aeroplane, to be overstressed is better than to be overtaken. We must accept with resignation and, if we can, with approval of its clean lines, a design of brain such that one in ten can never get home, even from a practice flight of fancy.

Perhaps the most intriguing, unexpected and essential of CORA's complexities is that three types of "memory" are demonstrably needed. The extension of the neutral stimulus is the first; but this is only in the nature of an after-discharge, not necessarily very different from the prolongation of the response seen in spinal reflexes, the effect that ensures that if you step on a tack, and your leg reflexly withdraws, it does not immediately come down again in the same place. (See Figures 12, 13, 14.) The second memory provides for the gradual accumulation of the separate overlaps between neutral and specific stimuli. This is on a much longer time scale; in CORA or an animal it might have to last for days or weeks, but must decline slowly in the intervals between experiences. In an electronic circuit a condenser is a convenient store; in the human brain some other mechanism may be far more compact and economical. A chemical change similar to the charging of an accumulator would serve, or a physical one such as microscopic growth at nerve endings.

The third memory is on a different scale again; in CORA it is an electrical oscillation at low frequency which subsides slowly over minutes or hours. This again is convenient for the model-maker, but does not necessarily represent the physiological truth. A swinging pendulum would have done the trick for the clockwork automata of Jaquet-Droz. In the brain, oscillating circuits have been imagined and some think identified; certainly we see oscillations in abundance in our

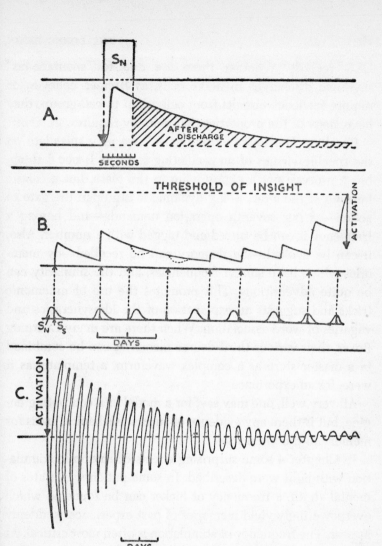

Figure 14. ". . . three types of 'memory' are needed." The Three Types of Memory. (a) Prolongation of the effect of a neutral stimulus equivalent to the after-discharge. (b) Summation of the combined effects of neutral and specific stimuli over a series of experiences. (c) Activation and preservation over a long period, of information about the significant coincidence between neutral and specific stimuli.

EEG records. Whether these are due to "spontaneous" rhythmic discharges in nerve cells, as Bremer believes, or require feedback circuits from cell-group to cell-group, they have many of the properties that memory requires.

Recalling Hartley's "vibratiuncles," we may speculate on the specific virtues of an oscillating store. It is not a thing, but a process; not a penny lying in the plate, but a candle burning at the altar. Being dynamic it can open the gate to action—as our seventh operation demands—and having a frequency it can be traced and tagged with a number. Also, it can be evoked—the memory can be recalled—by memories which have similar frequencies, and the similarity can be quite adventitious. This promotes the use of mnemonic tricks and suggests an explanation of the idiosyncrasies and vagaries of word association. When there are many Learning Boxes, their various third-degree memories can be combined in a master store as a complex waveform, a template, as it were, for an experience.

All very well, one may say, for a model or a computing engine, but is there any evidence of such a mechanical basis for memory in the brain itself?

In Chapter 4 some surprising effects of rhythmic stimulation with light were described. In some patients, in states of mental strain, a frequency of flicker can be found at which overpoweringly vivid memories of past experiences suddenly appear. The frequency of stimulation is often most critical. At 18 flashes per second perhaps the patient is overcome with a memory "as clear as crystal and as bitter as gall," and the brain is almost convulsed with electrical discharges; at 18.5 flashes per second the tempest abates; at 19 all is quiet. This can sometimes be repeated time and again, but usually, when

the mental state improves, the effect disappears. Repetition
is not always easy to test, because the patient remembers remembering and associates the experiment with the distress, so
that control passes from the physiologist to the psychiatrist.

No doubt in some people some memories may be local and
material, or at least mechanical. The observations of Penfield
on the effect of stimulating the brain during operations are
intriguing: when certain small regions of the temporal lobe
were stimulated some of his epileptic patients have been constrained to remember in vivid detail long scenes, tunes, acquaintances' voices, and to go on remembering them again
and again. But when this part is removed, the memory is
not amputated; which suggests again that memory is not a
place but a process.

For those who are repelled by the spurious magic of the
number seven, we may postulate, though we cannot prove,
an eighth operation in the learning process responsible for
the lifelong storage of the unforgotten and perhaps unfor-
gettable memories. In a model this would be similar to the
latching "memory" of *M. labyrinthea*, though in the living
brain a chemical lock would be more elegant. We must suppose that the gate would be wedged permanently open after
repeated confirmation of the hypothesis preserved by the
sixth operation.

The parts of the brain from which memories are evoked so
easily and regularly are those we find most liable to exaggerated electrical discharge during flicker, and it is here too
that in normal subjects the pattern of incoming stimuli can
be seen abstracted and preserved for some time after the
stimulation has ceased.

Multiplication of Learning Boxes has one particularly im-

portant consequence which is fundamental to our notions of brain design and to interpretation of electrical records from the living brain. The scheme of learning elaborated here involves two main groups of operations, one selective and the other constructive. In the latter group, the change of state induced by a series of observed coincidences in no way resembles the coincidences themselves; it is a formal, symbolic change, a signal of signals. When several such mechanisms are operating together in parallel and in series a new aspect of the constructive process emerges—abstraction. The several learning circuits are really extracting from a selection of events the features that are common to them in time and space— they are, in effect, *recognising a pattern*. It was found in a previous chapter that pattern is hard to define except as something memorable; here, starting from a different standpoint, we have reached the same conclusion—the raw material of learned behaviour is symbolic abstracted pattern. Turn again to the diagrams in Figure 13. What becomes significant, after statistical selection and constructive preservation, may be very far removed from the nature of the original incoming signals; it is a private image of their relations to one another —an idea.

A natural question at this stage is: Are we discussing what has been observed in these models or what may be happening in a living brain? Of course we are considering both, hoping that the explicit clarity of the first will illuminate the implicit obscurity of the second. Figure 15 shows how the seven operations of learning, inferred from first principles, could be performed by an assembly of nerve cells. Four types of nerve junction must be assumed, and junctions with the necessary properties are familiar. The electronic circuit which would be the precise equivalent of this neuronic system is given in Appendix C.

From study of the models, then, we can predict what a living system should look like from the outside while receiving and operating upon incoming signals. Since we designed

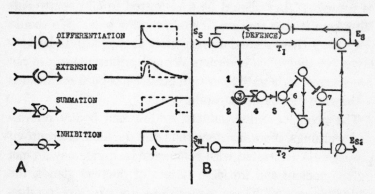

Figure 15. ". . . how the seven operations of learning could be performed by an assembly of nerve cells." (a) Conventional signs for indicating the four types of nerve junction necessary for learning, used in diagram (b). (b) The simplest neurone circuit which could perform the seven operations of learning.

the model to reproduce the features of animal learning it is not surprising that there are in fact superficial resemblances. But are there also more intimate subtle similarities? Have the electrical changes in the brain during stimulation any properties that our models also have? First, the statistical selective operations demand that every signal entering the brain should be turned into two derivatives, one clipped and shortened with respect to time, a signal of change and beginning, the other also clipped so as to ignore continuity but stretched at the same time, outlasting the signal if it is brief. The first derivative is important in case the signal turns out to be implied by something else, the other is important in case it implies something. Both these forms should be found far from the brain receiving zones where the signals first arrive—indeed they should reach nearly every region, ready to be mixed with any other signal with which they may tend to coincide. They should be transmitted to all departments "to whom it may concern, for information only." We should not expect these secondary responses in the brain to be very large, for they have low physiological priority; they are not even memoranda until their significance has been established —they are merely propaganda.

The need for indiscriminate, widespread bodily reforms and warnings, more insistent than the detailed intermittent propaganda associated with systematic learning, may be met by the ancient and intricate system of ductless glands, the endocrine system. These mechanisms are *effectors*, producing, instead of muscular movements, chemical messengers, the hormones, which circulating in the blood-stream initiate and control many processes connected with growth, metabolism, sexual activity and preparations for fight or flight.

One of the best ways of getting miners quickly out of a dangerous mine is to break a stink-bomb in the air intake to the ventilation pumps. Similarly in the body the nervous system can revolutionise in a few seconds the whole internal economy of the body by activation of the adrenal gland, evoking as it does so feelings of anxiety and tension as the organs affected are mobilised for action and signal their readiness back to the control. As already suggested in connection with long-term memory, specific chemical changes may well play a more subtle and discriminatory part in the reinforcement of the propaganda policy worked out by the learning mechanisms in relation to their experience.

Until the introduction of sensitive and discriminating electrical methods of analysis the search for physical evidence of these signals would have been vain. Experiments on anaesthetised animals are for this purpose of course quite futile—anaesthesia means not perceiving; and the human EEG is so attenuated by the skull and other intervening tissues that we can detect the electrical behaviour only of large clumps of brain-cells with ordinary records. But the combination of rhythmic stimulation by flicker, with automatic and toposcopic analysis of frequency and location, extends the resolution of the recording system by a factor of ten or even a hundred. Actually, such an apparatus has properties not unlike those of the brain itself in that it distinguishes more clearly between regularity and randomness.

When we started to analyse records from subjects undergoing stimulation by flickering light we were at once struck by the complexity and dispersion of the evoked electrical changes. The character of these responses and their relation to the subjective feelings of the subjects have already been described; their significance puzzled us for some time and was rather neglected in favour of the more dramatic and clinically valuable effects produced in epileptics by the same procedure. As long as we were restricted to analysis of frequencies in conventional records we felt some lingering doubt about the reality of these widespread and variable evoked responses. In certain conditions, frequency analysis can give misleading or at least ambiguous results. However, when toposcopic methods came into operation we quickly realised that the secondary responses have a very real existence and

must be accepted as a significant though variable feature of
normal brain physiology. An experiment which can never be
repeated does not appeal to the laboratory experimenter, but
it slowly dawned on us that the fluctuations we were inclined
to dismiss as inordinate experimental error were the heart and
substance of the remote effects of rhythmic stimuli. Again
and again during a toposcopic study, the screens would light
up in a brilliant pattern of interlaced gleaming threads as the
flicker was turned on; then, when we had adjusted our controls and prepared to photograph the display—all would become drab and inconspicuous. Sometimes again the startling
effects would come at the end of the experiments, when the
subject was feeling tired and uncomfortable. No two people
behaved in the same way, nor did any one subject give quite
the same picture on two occasions.

Careful study of such records as we did succeed in snatching from these momentary conflagrations showed that there
were at least two effects. One, most prominent at the start of
the stimulus, was in the nature of an "on effect," a brief spike
appearing in many areas, a disseminated transient. The other
was more complex and long-lasting but equally widespread—
a train of diminishing waves, like the rumble of thunder after
a lightning flash. Similar effects have been seen in experiments on animals but in more restricted distribution; the response even of a primary receiving area is neither simple nor
regular. In a conscious normal human subject the secondary
derived components may overshadow the primary elementary ones—but only for a time.

It was partly in order to account for these observations that
the analysis of learning was undertaken; not until the analysis
was complete and its implication was realised, however, did

an explanation of them present itself. These evanescent, widely disseminated responses are of course just what our theory of learning requires: their transitory quality is a sign of the insignificance, the meaninglessness of the flicker itself. When the stimulus is first presented, as far as the subject is aware it might imply anything—or be implied by anything; but as flash succeeds flash in monotonous series, the possibility of significance is dismissed. That is, unless the whole situation takes on a peculiar character—of pleasure or discomfort. Then the response may augment and spread and the subject complain that "the light makes my head ache," drawing a reasonable conclusion of causality which experiment shows to be false. But the relation between repeated coincidence and causation must be discussed later.

Consideration of normal flicker responses in these terms led to a renewed study of the exaggerated and apparently anomalous effects found in many epileptics and some normal subjects. At first, it seemed that these evoked discharges should be a perfect testing ground of the learning theory, since they represent in the brain an event of gigantic magnitude, a bomb to a squib as compared with the ordinary effects. But these electrical experiences, however impressive their scale, proved astonishingly intractable to conditioning. It was only in one or two subjects that a loud sound, regularly repeated just before the flicker was started, would eventually evoke or facilitate an unusual response, provided that the original flicker response was an exaggeration of only one normal feature. Those subjects who displayed both fast and slow components to excess during flicker might never have heard of Pavlov or Yoga.

Considering how easily obscure and unobtrusive bodily

functions can be conditioned, it was disappointing that a major event in the brain itself was so inaccessible. But an explanation now seems fairly obvious. The abnormal effects may be signs of disorder or weakness in the parts of the brain responsible for the third and fourth operations of learning, the mixing and summation of specific and neutral stimuli. If this is so, naturally the one physiological activity that cannot be conditioned is a fault in the conditioning mechanism. You cannot learn not to learn.

CHAPTER 8

Intimations of Personality

Children use the fist Until they are of age to use the brain.
Elizabeth Barrett Browning

THE PHYSIOLOGIST of a generation ago, baffled by the complexity of brain mechanisms, would have been doubly reluctant to enter the territory now reached in the study of them. Those were the days when the specialist was in the ascendant and won golden opinions precisely by keeping to his own field, days before physician and psychologist put their Jovian heads together to produce the psychiatrist. The physiologist of that period was a shy fellow, having indeed little to contribute to the Minervan synthesis. He was more concerned to retain and advance the claims of physiology as an exact science, content to be the servant of the practitioner rather than his partner. Consequently his public repute remained similar to that of the physicist in the days when physics only came into the school curriculum as a stray half-hour of "Heat, Light and Sound."

So far this account of the brain has been traditional in scope if not in matter; that is, brain has been considered in general terms, as might have been heart or liver or any other organ, as if it were anybody's brain, apart from some passing references to typical variations. This common character of an organ and its functions is a natural assumption of the physiologist, whose business it primarily is to establish a norm for the pathologist rather than explore idiosyncrasies. But brain is the organ of personality, and enough has been said in connection with the individuality of the alpha rhythms and the variable experiences of the subject in the flicker test, to indicate that at some point we should have to begin to examine phenomena of the living brain as elements or at least intimations of personality.

The more we learn about the brain, the more clearly we see that it can only be studied profitably as a complex of mechanisms; and to study all the mechanisms of the brain implies reference to all its functions, including phenomena which are the physical counterparts of mental events. Thus if the brain physiologist today passes across an old scientific frontier, it is not to join—though still serving—the practitioners, but as an entirely fresh approach to the study of the everyday working of the human mind in so far as it is capable of quantitative observation. To some psychologists this appears to be undermining their position; by others it is welcomed as possibly providing some elements of a more secure material basis for a science that has neither units nor axioms.

Still more consequential in time may be the arrival of the brain physiologist in the social field. He is already providing a certain amount of public service, but not one thousandth part of the assistance he is likely to be called upon to give when "prevention is better than cure" becomes the watchword in mental as well as physical health policy.

It was in fact in connection with clinical service that the first information of specific importance concerning mechanistic intimations of personality was discovered. This was during the war period in the EEG laboratory of the Burden

Neurological Institute. It will be recalled that one of the first curiosities which had been noticed about the alpha rhythms by early electroencephalographers had been that in no two people are they the same. Not even identical or uniovular twins have quite identical alpha patterns. But in any one person the pattern is remarkably constant from year to year, once cerebral maturity has been reached at the age of 14 or so. It was noted that the differences between individuals are very great. Early speculations did not go beyond that; the differences were regarded as otherwise meaningless.

The individuality of EEG records, brainprints, seemed to be similar to that of fingerprints, giving simple identification. This in itself was a physiological novelty. The skin is individualist also to the point of rejecting a graft of any other person's skin, but no organ was known to furnish positive identification by its behaviour. And later observations in a wider field showed there was more in it than that. When the correlation between certain phases of rhythmic activity and certain features of personality was first demonstrated, it was as surprising as if the loops and whorls of fingerprints had acquired meaning. The main facts which have since been verified in this connexion can now conveniently be presented in sequence of age.

Among the first papers published by Berger was one on the relation between age and the EEG. The observations he had made with his too simple equipment suggested that little electrical activity of any sort was present up to the age of one month, and from this age, up to a few years old, amplitude and frequency increased steadily. His observations were generally and broadly confirmed. It soon became evident, however, that some activity is present from birth, and that its increase is not simply a function of age. It has now been found that even before birth some electrical activity can be detected.

If electrodes are attached to the belly of a pregnant woman in the eighth month of gestation, whenever the head of the child moves into the region near them, slow irregular delta waves can be recorded. Sometimes the delta activity is interrupted by larger and more rhythmic discharges similar to the wave-and-spike patterns usually associated with epileptic attacks. It is unlikely that all the babes in whom these patterns have been detected before birth will turn out to be epileptic; rather we may suppose that, while nestling in its private pool within the womb, the child approaching full term is outgrowing the resources of that haven. The convulsive twitching and stretching of the unborn child is evidence that its oxygen supply is lagging behind its needs; with growth the deficit increases, and with it the petty seizures; until, at the appointed phase of some maternal tide, half suffocated, the baby thrashes its way to freedom or disaster. So, too, the pulsations of a jellyfish, augmented in the breathless oily calm of a summer sea, drive it toward regeneration in the foam of breakers.

At birth, and for some months after, the main feature of the EEG is still irregular delta rhythms. The more passive and somnolent the infant, the more prominent are the delta rhythms. Even in the first few days of life there is a marked difference between the sleeping and waking patterns of EEG. Changes of frequency are correlated best with brain weight, changes of amplitude with the number of active neurones in the first few months and with skull thickness thereafter.

For some time it was thought that the alpha rhythms of

adult life are rarely found in children less than 8 years old; later analytical technique was more revealing, and a much greater number and variety of childhood rhythms have been identified. (Figure 16.) We can now say that no records of children below the age of three are found which could be accepted as normal by adult standards. From that age, however, the typical features of the normal adult record appear more and more frequently, and even below the 3–4 years age-group some alpha components are found—small, diffuse, unresponsive, and usually masked in the primary traces of the record. Rhythms of the alpha type then begin to appear in short bursts, often at considerable amplitude. In some children of 4 they are even the dominant feature of the record, although the typically infantile rhythms, both delta and theta, are still clearly visible.

In most children the records vacillate between the theta and alpha types for some years, at least until the age of 10 or 11. It is noteworthy that a record with no rhythmic activity at all—common enough in adults—is extremely rare in children below the age of 12. When the alpha rhythms first appear in very early childhood, they are scarcely responsive even to arresting visual stimulation; they begin to show an unmistakable connexion with vision only after the age of 3–4 years. The effect of mental activity or non-visual stimuli cannot be very accurately tested in infants, but the classical blocking reaction appears clearly at the age of 6–7 in some children. Statistically adult responsiveness is not found before the age of 10–11, when also some children show a very low amplitude rhythm even with the eyes shut, and others a persistent one, as in the adult population.

Absence of alpha rhythms in adults is normally a sign of

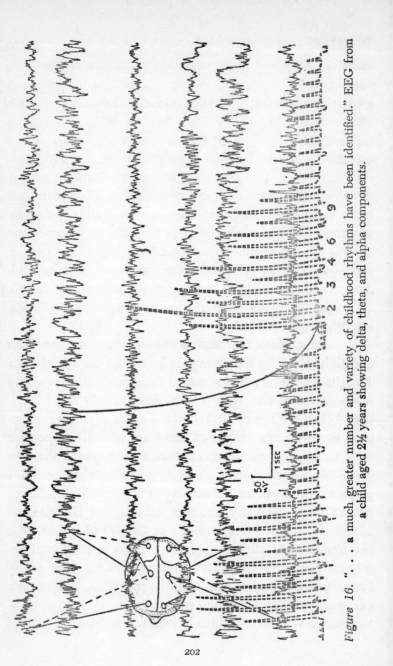

Figure 16. ". . . a much greater number and variety of childhood rhythms have been identified." EEG from a child aged 2½ years showing delta, theta, and alpha components.

50 mV
1 SEC

1 2 3 4 6 9

vivid visual imagination; in babies and young children this seems an unlikely explanation. Judged from their behaviour, children are relatively unimaginative below the age of 3 or 4. So the later appearance of alpha rhythms and their relative unresponsiveness may be a sign of the maturing faculty of visual imagination and the lack of practice in its use.

There will be more to say about delta later, essentially the billowy rhythm of sleep; nor is it surprising to find that in their electrical behaviour the brains of infants and adults resemble one another only in sleep. Sleeping like a child, we say. But also, behaving like a child. In the first case our brain is rocking gently on the delta waves of the cradle; the other is not so simple, though the saying is equally apposite.

Apart from the lowest frequencies in the juvenile spectrum, probably associated with the earliest and most vulnerable stages of cortical development, the most arresting feature in the analysis of children's records is the activity at frequencies between 4 and 7 cycles per second. These rhythms were still regarded as merely transitional between the very slow rhythms of the infant and the faster alpha activity of the adult when chance observation of the subject's change of mood directed our attention to the possibility of a more significant psychological correlation. With further observation, the following points among others were noted. Theta activity is usually dominant in records of the 2–5 years age-groups; it is approximately equal to alpha activity at 5–6 years; it is smaller above 6 years, and very small or intermittent at 10. With other technical observations, a case was thus made for regarding this group of rhythms as specific; they were named "theta" because they were first identified in clinical studies as arising in or near the thalamus, the antique bridal chamber of the brain, considered by some as the factory of feeling, the seat of emotional display.

Once mechanically distinguished in this way, there was little difficulty in enlarging upon the first indication that theta activity was connected with the mood of the subject. No marked response was shown to ordinary visual stimulation, as with alpha activity, but it soon became clear that theta is associated in some delicate fashion with pleasure and pain. It is easily evoked in a young child, for instance, by frustration, by snatching away a proffered sweet. But the change may be objectively apparent as pleasure (cooing or smiling) as well as annoyance (whimpering and kicking).

Even in earliest days, and throughout childhood, individual differences are striking. In some children the theta activity is clearly associated with affective changes of any sort; in some it is linked only with pleasant feelings; in others only with unpleasant ones. It is a truism that in a child below the age of 10 years or so the show of feelings is more exuberant and less intellectual than in older people. It may well be that this characteristic is associated with the late development of certain of the cortico-thalamic connexions and the consequent relative dominance of thalamic activity (with which the theta rhythms are indirectly associated) over behaviour during the first years of life. Such relationship, easier to assume than to establish, unless a Slaughter of the Innocents is contemplated, is at least a useful hypothesis.

The connexion between the changing of the rhythms from age to age and the anatomical maturing of the brain is evidently complex. The extent and nature of these changes, we may assume, will depend upon which of the innumerable areas and connexions of the brain are mature; and the dependence will not be a simple one; for the changes will be affected not only by the development of the nervous system but also by the presence of a complex private system of acquired responses and conditioned reflexes. Young children live in a world which few of them can describe and few adults remember; it is the special features of this world which, as much as anything else, determine the theta patterns of the young.

The notion of anatomical connexion with the theta rhythms is still more suggestive when we come to consider what is known of the phenomenon in conditions other than childhood. For some time "slowing of the alpha rhythm," or "deceleration of the dominant," had been regarded as an empirically pathological sign associated with organic lesions and psychiatric disorders. Its real significance appeared only when analytical unmasking showed that the appearance of "slowing" was due to the presence of a separate, slower rhythm, masked by or masking the alpha rhythms according to its relative amplitude.

The resulting correlation of the theta rhythm with the thalamus and structures around the third ventricle has been elaborated by Denis Hill in connexion with aggressive psychopathy. This led him in the course of further observations to recognise the existence of a group of psychopaths with a special characteristic unilateral slow theta discharge, which he termed "dysrhythmic aggressive behaviour" cases, or DAB. The type of behaviour characteristic of this group involved violent attacks on living creatures in the attempt, as Freud put it, to turn living matter back into the inorganic state. These destructive and murderous episodes were often almost or completely unmotivated by ordinary standards; they recall irresistibly the purposeless destruction tolerated but deprecated in young children when frustrated, or, to readers of St. Augustine, his angry baby whose "weakness of infant limbs, not its will, is its innocence."

St. Augustine would have been interested in the measure in which EEG records confirm the connexion between the sinful mood of infant and adult. Few of us indeed escape unscathed from the test. We are all miserable sinners. In ordinary circumstances the theta rhythms are scarcely visible in good-tempered adults, but they may be evoked even in them by a really disagreeable stimulus. It is difficult to arrange this in a laboratory, where its artificiality is as obvious to the subject as it is embarrassing to the experimenter. The subject must feel himself really deeply offended by some personal affront; one's enemies are not likely to yield themselves so conveniently into one's hands; of strangers one cannot know enough to be sufficiently offensive; and a friend one may not, by definition, offend.

For some years research itself was frustrated by this situation; our hypothesis predicted that theta rhythm should be evoked by suitable emotional stress in any normal adult person, but we could not bring ourselves to jeopardise impersonal relations with experimental subjects. We then had the notion of testing the effects of pleasant instead of unpleasant stimuli. In one of the earliest experiments, records were taken and analysed from a French student while his head was being stroked by a young lady from Wales. The stroking had no detectable effect upon the analysis, but we noticed that whenever the stroking was interrupted, a few seconds later the analyser indicated a sudden transient outburst of theta rhythms at 6 cycles per second. This suggested that the with-drawal of a mildly pleasant sensation was more upsetting than the administration of mildly unpleasant ones such as we had tried with this subject. Animal psychologists also have noticed the startling effects of mild pleasure ending. Hebb describes how a chimpanzee may be quite content to stare quietly for hours at an attractive female some cages away, then break down in a paroxysm of rage and exaspera-tion when she retires to her sleeping-chamber.

Short of strip-tease, we proceeded to tantalise a number of normal and clinical subjects by various innocuous devices. The responses varied with the subjects; but we discovered in many of them a remarkable property of theta activity—an extreme constancy of distribution and development of it in certain subjects throughout a whole series of experimental frustrations. In individuals of this constant sensitivity, when-ever a pleasant situation comes to an end, a theta rhythm be-gins to appear regularly a few seconds later and culminates in about 10 seconds, when it abruptly disappears. This cre-scendo, with its sudden finale, is so stereotyped that half a dozen such records, taken in a series of individual tests, can be superimposed on one another and even the details of the rhythmic discharge can be seen to coincide time after time. (See Figure 17.)

Why, in contrast to all other functions of the normal brain, should this particular response to cessation or lack of pleasure be so invariable? This still perplexes; it can only be conjec-tured that perhaps adjustment to frustration and disappoint-ment is one of the first and firmest foundations of personality. Just as, in learning, many possible ideas must be jettisoned in

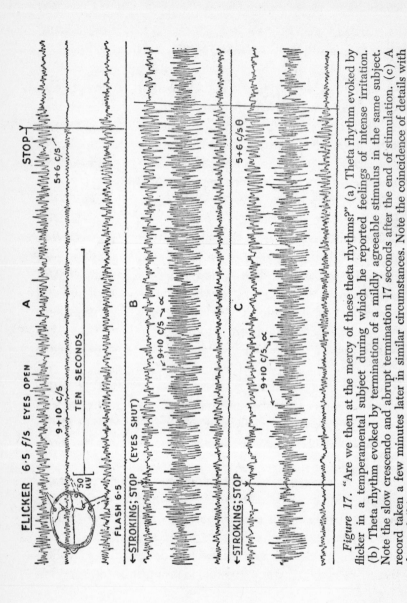

Figure 17. "Are we then at the mercy of these theta rhythms?" (a) Theta rhythm evoked by flicker in a temperamental subject during which he reported feelings of intense irritation. (b) Theta rhythm evoked by termination of a mildly agreeable stimulus in the same subject. Note the slow crescendo and abrupt termination 17 seconds after the end of stimulation. (c) A record taken a few minutes later in similar circumstances. Note the coincidence of details with those of (b).

208

the search for significance, so there may be more unpleasant ideas than pleasant ones, and most pleasures are indeed but fleeting.

In an earlier chapter some evidence that the alpha rhythms provide a scanning mechanism for visual signals was considered. If the alpha rhythms scan for pattern, we may perhaps consider the theta rhythms as scanning for pleasure. On this basis the uniform swell of theta rhythms, as pleasure fades, would represent the renewed search for other pleasures, while its abrupt snuffing-out is a measure of the individual's habit of stoical acceptance and resignation, an admirable and hard-won fortification against vain regret. As a corollary, the persistence of theta rhythms in children and ill-humoured adults may be a simple sign of inexperience in the first, of intemperance in the second.

In bad-tempered adults, especially in those with an unusual tendency to aggressive behaviour, the theta rhythms are often prominent and may sweep through quite a large area of the brain. Their childish intolerance, selfishness, impatience and suspicion are mirrored in the juvenile appearance of their brain patterns. Are they still children at heart? The coincidence is so obvious that we are tempted to jump to the conclusion that the theta rhythm of childhood is qualitatively and functionally identical with—as it is quantitatively similar to—the theta rhythm of adult pathology, sign of relative immaturity of the mechanisms linking cortex, thalamus and hypothalamus. How convenient it would be, and socially how useful, if one could classify the whole population morally by examining their theta rhythms! But the case is not so simple. It seldom is. As so often in scientific research, when a goal seems to be in sight you find something quite unexpected round the corner. In this case it was, instead of theta rhythms, delta.

With the assistance of a public authority it was possible to examine a considerable number of intellectually normal delinquent children and to relate the features of their EEG's to an assessment of personal and social backgrounds. The odds seemed to be heavily in favour of finding pronounced theta activity as a sign-manual of mischief. One might have expected to be able to pick out the really "bad boys" by their theta signatures. Actually 85% of the records were "abnormal" or at least peculiar, but no significant correspondence was found between the presence of theta rhythms and particularly bad behaviour records. The whole research seemed to be otherwise negative until, after tedious statistical tests of the association between other features of behaviour and the results of automatic analysis of the EEG records, it was found that the significant rhythms in 70% of the children were not theta but the slower delta activity, hitherto associated only with pathological conditions, with infants and, as will be discussed in more detail later, with sleep. The only characteristic of behaviour which was significantly correlated with the presence of delta activity in these children was not aggressiveness, but a relatively *promising* reaction to their mothers, to leisure and to their fellows, as estimated by the authorities responsible for classifying them.

Consideration of what these estimates imply has suggested that the common factor related statistically to delta rhythms is a comparatively docile attitude to suggestions from others. The terms "malleable," "easily helped," "easily led" were used, and the word that seems most apt and free from irrelevant or misleading associations is "ductile." The significant delta activity in these boys was small and hard to measure, even with the best analytical equipment; it was often masked completely in the primary records by alpha or theta rhythms; but the suggestion of something infantile is hard to escape. The ages ranged from 10 to 17 and due allowance was made for the normal differences between juvenile and adult records. There was no correlation with age, so arrested or delayed maturity seems an inadequate explanation. The delta distribution was not the same as in babies; it was mainly in the temporal, parietal and occipital lobes, but usually diffusely located; and in only a few boys could it be compared directly with a baby's record. There is doubtless more to be learned from this observation and particularly its relation to the various contingencies of education and social rehabilitation.

Evidence that the condition of the electrical activities—alpha, delta and theta—is closely connected with the maturing of the personality is plentiful. We should expect therefore to find signs of mechanistic immaturity in cases of disturbed behaviour. By watching the decline or development of brain rhythms in a child it is possible to follow to some extent the formation and growth of character, and so be guided in or warned against correction or encouragement.

The correlation between deep feelings, the wells of personality, and the apparently simple electrical discharges which we record and analyse, may seem remote to those who have never witnessed the connexion between them either as the result of statistical investigation or in the course of a laboratory experiment. A few minutes of theta flicker would probably be convincing. Visual stimulation at the frequency of the theta rhythm evokes, even in a normal subject, a feeling of annoyance and frustration. If you are an irascible type it may make you very angry. The mood changes completely without any external emotional stimulus. If, however, during visual stimulation, an emotional stimulus is added— if the subject is told something really disturbing or annoying to him—the effect of this double stimulus is a summation of the two effects of flicker and emotional stimulus when given separately. (See Figure 17.)

Are we then at the mercy of these theta rhythms? Do they provide an excuse for outbursts of uncontrollable bad temper in young and old? By no means. Laboratory experiments confirm most of the admonitions about learning to keep our temper, which we heed when young or wish we had in later years. The characteristic sign of those who have learned to do so can be recognised in the laboratory. When the theta rhythm recedes, after its first evocation by flicker, we are getting a quantitative description of the subject's efforts to suppress the bad feelings that accompany it. These coded messages from his brain can be deciphered, following the ups and downs of the conflict, as he struggles to keep his temper or to dissipate feelings of annoyance or frustration.

There is a simple proof of the validity of this decoding: if the flicker stimulus is given in equable emotional conditions and the subject is encouraged to control the feeling of annoyance aroused by it, the evoked electrical disturbance is quenched—he suppresses his bad feelings, and with them their material associates, the theta rhythms. In this experiment, easily repeated in any laboratory equipped for flicker and analysis recording, there can be no doubt that we are witnessing the physical counterparts of mental events.

Here, for once, when the coded signal is deciphered, the

message turns out to be a very personal one; also the information it contains can be verified through other channels, by
the subject's account of his experience. We still do not know
enough about the delta and theta rhythms to say much more
than this; but, given a wider field of observation and a more
varied correlation of the rhythms with personal characteristics, there is promise of much information of potential social
value. Already, as indicated, some public use is being made
of the assessment of character which statistical and analytical
correlation of the recorded rhythms can contribute to the
general picture; and the forensic use of EEG is now orthodox
in criminal procedure.

Discussion here has been limited to aspects of the rhythms
dominant in childhood about which interpretation could be
offered with some degree of confidence. There are plenty of
hints of other significant correlations to be discovered by
patient work in this field, not only in the characteristic
rhythms of the child but in the development of the adult alpha rhythms and their relation to the maturing character of
the brain. An index of special aptitudes and ineptitudes, of
temperamental maturity, tendencies of self-indulgence and
self-control, as shown by the mechanistic responses of the
brain, for which there can be no coaching, will be possible in
the near future. It may prove to be a useful supplement to
the conventional IQ and personality tests.

The alpha rhythms, as might be expected, are a source of
more general information about personality than those of
childhood. The irregularity of their first appearance has been
noted—they emerge sporadically even below the age of 3
and become dominant in rare cases at the age of 4, although
typically adult responsiveness is not generally found until

about the eleventh year. The differences are so wide as to seem random, but after our experience with first impressions of "the Berger rhythm" we must hesitate to conclude that they are meaningless. About the brain it would be better to assume that nothing means nothing. Some of the records of children being made today will be interesting reading when examined in connexion with the adult achievements of those children. Meanwhile there is much to be learnt today from the variations of mature alpha rhythms.

Meaning began to emerge from what had seemed random differences of personal rhythms as soon as a sufficient number of records had been taken to show that there is a natural grouping of differentiated responses to the normal blocking by mental effort. It has already been mentioned as one of the earliest observations of EEG, that in most people the alpha rhythms, prominent when the eyes are shut and the mind is at rest, disappear whenever the eyes are opened or when the subject makes a mental effort—for example, while doing a sum in mental arithmetic. Exceptions had of course been noticed; entire absence of rhythms in some cases. But it was only in the course of war services at the Burden Neurological Institute that we were able to designate some of these exceptions as a stable group with definite characteristics. It was shown in 1943 that individuals with persistent alpha rhythms which are hard to block with mental effort, tend to auditory, kinaesthetic or tactile perceptions rather than visual imagery. In this group of persons the alpha rhythms continue even when the eyes are open and the mind is active or alert.

The group with persistent activity is known as P for short, while the larger group, whose alpha rhythms are responsive, are known as R. A third group was further definable as those people in whose EEG's no significant alpha rhythms are found, even when taken with the eyes shut and the mind idle. This group is known as M, for minus, and consists of persons whose thinking processes are conducted almost entirely in terms of visual imagery. (See Figure 18.)

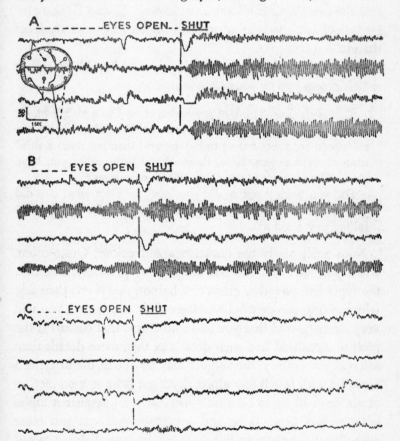

Figure 18. ". . . a discrepancy in their ways of thinking." Alpha Personality Records. (a) R type. (b) P type. (c) M type.

Several surveys have been made to find out how these types are distributed in the population; groups comprising more than 600 persons have been studied for this purpose. The proportions vary a good deal according to occupation, but, in general, about two-thirds of an ordinary normal group of people selected at random are found to be of the R type, and the remaining third are about evenly M and P. The proportion of M types is usually rather higher in science students than in arts students.

Here is a simple test by which you may be able to tell to which group you or your friends belong:

> Shut your eyes. Think of a wooden cube like a child's block. It is painted. Now imagine that you cut it in halves across one side, then cut these halves in halves, and then cut them a third time at right angles. Now, think of the little cubes you have made. How many of their sides will be unpainted?
>
> Did you work it out or did you "see" it? Then what else did you see? What colour was the cube? Did you see the sawdust falling as you cut it?

Note particularly the question about colour. Quite often, when you have put the test to someone precisely as above, the reply will be red or green or whatnot; and if you then ask, why red or green or whatever the colour mentioned, you will very likely be told that you had said it was that colour! If the picture visualised had such details as this, more details than are really necessary, the subject may be one of the M type, a visualist with few if any alpha rhythms; if he saw no picture at all, he is likely to be a non-visualist with persistent alpha activity, a P type. If he had a picture that was just clear enough for the purpose but no more, he is probably a mixed type, mostly R, with a responsive alpha rhythm.

When a solution or decision of any kind can be reached by visualising it, the performance of the M type of people is rapid and precise; but when they are faced with a problem of an abstract kind, or one in which the mental pictures required are too elaborate for them, they become sluggish and confused. On the other hand, at the other extreme, the small P group, people whose alpha rhythms persist even with the eyes open and doing a sum, do not use visual images in their thinking unless they are obliged to do so. Even then, their mind's eye is almost blind; they think in abstract terms, or in sounds or movements; they may even have to "feel" their way out of an imaginary maze. The R group, the responsives, whose alpha rhythms disappear when they do a sum or open their eyes, are intermediate between the other two groups; while they do not habitually use private pictures for their everyday thinking, they can evoke satisfactory visual patterns when necessary. Moreover, they can combine data from the various sense organs more readily than can either the M or P types. So even if you did the cube test just now by visualisation, it is possible you belong to the more adaptable and versatile R group. The test of the cubes is valid as far as it goes, but only a recording and analysis of your EEG could give you your standing in your class.

There is still a great deal to be learnt about these groups. For instance, the origin of the differences between them. When and how does the differentiation become so clearly defined? Why are the alpha rhythms so persistent when they first appear in childhood? We must be cautious about jumping to any such conclusion as that children only learn later to think in visual terms, although this is suggested by the extreme rarity of M type children. If, as seems likely, imaginative thinking becomes habitual at about the age when alpha rhythms appear, the startling differences between children, and the critical influence of age on the effects of deprivation, may find an explanation in the tardy and variable development of these physiological mechanisms, which in the adult provide the consolation in exile so conspicuously lacking in the very young. We have not yet had time to follow the development of a large enough population from birth to maturity to discover how soon and how permanently these differences are established.

Evidence already available, however, both statistical and experimental, strongly suggests that the alpha rhythm characters are inborn and probably hereditary. We are also beginning to get some notion of the distribution of rhythm characteristics in the population. Alpha rhythms vary in frequency, as we have seen, from 8 to 13 cycles per second. The distribution of these frequencies in the population seems to be as normal as the variations of stature; so that, like stature, it may well depend on many factors, some hereditary and some environmental. But the way the alpha rhythms respond to mental effort, and the consequent grouping which has been described, is less straightforward in its distribution. It may be more like eye-colour or blood-group in its dependence on heredity—complicated, however, by the effects of individual experience and impressions.

Differentiation by experience can be most clearly traced in the case of identical twins. The resemblance between the alpha rhythms of uniovular twins is as close as that of their fingerprints; the resemblance of the unstimulated rhythms persists through the years. Differences soon begin to appear, however, in the details of their responses to stimulation. The similarity of their brain mechanisms may continue to be as close as their physical resemblance when they mature, but their conditioning experience will not have been precisely alike. The differences of these imposed patterns show not in the records of their resting rhythms but in those of their responses to stimulation. They may still seem identical to casual acquaintances, as much alike as ever in form and feature; but the experimental responses of their EEG's will indicate slight differences of character which possibly only their intimates will have noticed. When we find such evidence in our records, we seem to be justified in saying that we not only detect but are also able to measure acquired differences of personality.

An appreciation of theta activity has various applications and possible uses. In what way could this information about alpha rhythms be of service? There is no doubt that the different characters of the three personality groups—P, R, M—and the effects of their different ways of thinking are constantly intruding in our daily life. Everyone is familiar with the unaccountable nature of (other people's) family squabbles and perhaps even our own lovers' quarrels. Apart from arithmetical problems, in which nobody is much concerned about how they are done as long as the answer is correct, the three different ways of dealing with any question may, and often do, lead us far apart before we reach a common destination. Here is a simple example.

Supposing Peggy and Michael at breakfast receive an invitation to a party and have to decide whether they shall go to it; and supposing Peggy is an extreme P type and Michael an extreme M type. Michael will have a whole series of vivid pictures about it all in his mind; he will see them going to the party, the party itself, the people they will meet, and so forth; and he will compare it all with alternative pictures of what they will be doing if they do not go, perhaps a vision of somebody awkwardly asking him the day after it why they weren't there. That is the way M types literally figure things out for themselves. He will come to his decision that way very quickly, and in discussing the pros and cons with Peggy he will try to make her figure it out that way, too. But Peggy, being a P type who does not use visual images in this quick and easy way, has a more abstract method of thought, and will be considering the advantages and disadvantages of going, balancing duty and other obligations against the pleasure of an outing, the convenience of going against the effort, the number of times "we've been there without asking them here," and so on. She will be irritated by Michael's efforts to make her see his pictures, while he will be equally annoyed by her attempts to make him appreciate her heartfelt abstractions. It is not that one of them is more self-centered than the other, though if the reader happens to be of either type he may already have decided that the other one is. Worse than any adjustable blinkers of that kind, their language, their mental accents, so to say, are incompatible. Through nothing but the differences of their ways of thinking, before they can come to agreement, things may get to such a pitch that neither will give the other credit for clarity, consistency or good taste.

Fortunately extreme types are rare; but when two people display unreasonable and irreconcilable differences of approach to a question, before concluding that this is due to innate antagonism or incompatibility of purpose, a discrepancy in their ways of thinking may be worth looking into.

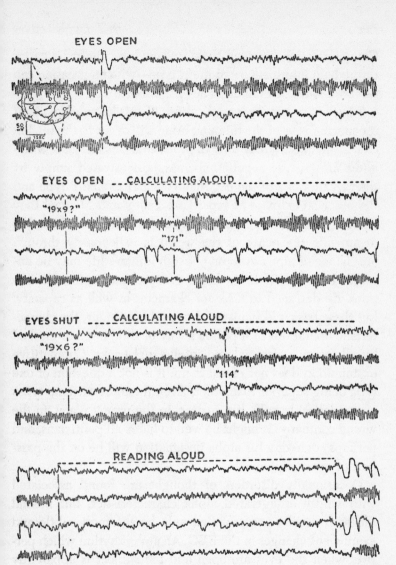

EYES OPEN

EYES OPEN — CALCULATING ALOUD

"19 × 9 ?"

"171"

EYES SHUT — CALCULATING ALOUD

"19 × 6 ?"

"114"

READING ALOUD

Figure 19. "Fortunately extreme types are rare." Record of Extreme P Type. So persistent were the alpha rhythms in this subject (Peggy in the fable) that the only way to stop them was to make her read aloud.

Communication between them, meanwhile, might be easier through an intermediary of the R type—who on occasion can use both ways of thinking. (See Figure 19.)

Suggestions for other applications of this knowledge about the mechanisms of the thinking brain will occur to the reader. One is obvious in these days of protracted international wrangling. How many negotiations may be frustrated simply by the fact that one of the negotiators is an extreme P type and the other an extreme M type! Like Peggy and Michael they want to agree and eventually may come to an agreement, but meanwhile the peace of the world is in jeopardy through mischance of alpha grouping, just as a man's life may be imperilled by a mistake in blood-groups. Academic examinations are designed to discover character as well as capacity, but these basic mechanisms of mental behaviour, the characteristic operations of a person's way of thinking, are masked by all manner of social and intellectual tricks. Competitive examination does not reveal them; it is not a question of one type being superior to another. Even the most extreme types are undesirable only, like matter, in the wrong place, in the wrong company. It might be well to index all politicians, and perhaps some day his alpha designation will be on the passport of every diplomat.

Occasionally disorders of thought are found associated with wildly exaggerated alpha characteristics, but mental illness is usually accompanied only by the most subtle and evanescent changes in the EEG. An alpha rhythm which persists when the eyes are open and the subject is apparently fully occupied—reading aloud for instance—is usually suggestive of some isolation from reality. In a few cases absurdly persistent alpha rhythms have been the first clear indication of something wrong, that what seemed unintelligible or eccentric brilliance was really lunatic delusion.

Rhythmic activity in the alpha range of frequency at 9 to 10 cycles per second is sometimes found in the front rather than the back of the head, with the activity in the two hemispheres seemingly quite dissociated. In such cases the "alphoid" rhythm, as Sessions Hodge has dubbed it, is not reduced by visual or mental activity. A significant proportion of such subjects are referred for examination because they have suddenly committed some offence in an acute confusion of lust and frustration. In some cases, when the confused episode is over, the alphoid rhythm disappears and a normal occipital alpha rhythm may be found—at a different frequency from the frontal one. These are truly miserable sinners, as compared with the insane with persistent occipital alpha rhythms, who often seem quite content in the private circle of their intractable fantasies.

Generally, then, the electrical signs of brain function correlate better with "how" than with "how much." The superficial EEG differences are linked more closely with personality than with intelligence. There is one measure, however, which seems to vary with extremes of dullness and brilliance. Soon after we started making regular frequency analyses of records it was realised that the results depended to some variable extent on the length of time over which the analysis was made. In the standard instrument the analysis "epoch" was ten seconds; if we added up the readings over six such epochs and divided each of the 144 readings by six we obtained an average frequency spectrum for one minute. This average sometimes looked very like the separate analyses from which it was made up; sometimes it was quite different. The greatest variance we found in our more brilliant colleagues or friends, the least in our duller patients. Revolting at the prospect of so much arithmetic, we devised a computing circuit to do the job automatically; nearly all analysers are now fitted with this "averager" and various lengths of periods can be averaged. With this device it is possible to extract a measure of what one may call the "versatility" of the brain.

The more original types, verging towards genius, seem to have a high versatility in this sense; to obtain a set of average results similar to one another, one has to average over a very long time—several minutes, at least. In "dull" brains half a dozen ten-second analyses may be indistinguishable from one another.

This may seem a difficult statistic to understand, but in fact one uses similar judgments every day—without arithmetic. A "brilliant" driver, for example, is one who goes very fast when speed is safe, very slowly when caution is indicated, who accelerates briskly and slows down quickly, matching his performance to the conditions of the road. The average speed of such a driver, taken over periods of, say, one minute, would show great variations, and only his lap times would be constant. But a dull driver creeps round at a low steady average; his speeds on bends and on the straight are not very different, and they compare with his lap speed, too. So the brain seems to vary in the span of its variance, matching in the brighter ones the scale of its effort to the quality of the task.

In an ageing population, attention naturally veers toward the art of growing old. The brain is not, in general, the limiting factor in determining the length of our days. Disorders of the heart and circulation, the appearance of malignant growth, accidents and social isolation, are responsible for the greater proportion of "old age woes." There are of course certain brain diseases characteristic of the last few decades of the extending normal span, but these are relatively rare and specific disorders. The EEG changes little with advancing years; apart from truly senile states, it may have the same features at 80 as 60. Gerontology, the science of old age, has so far gained from electrophysiology only the assurance that most brains could outlast the other organs.

The dying brain is calm; as the blood reaching it brings less and less oxygen, a few slow waves appear with failing awareness; these rise in amplitude, then slowly wane, and with them fades the organisation of personality. Strangely, the brain and its subordinate ganglia are ill-armed against oxygen lack. A man dying of coal-gas poisoning or exposure to a rarefied atmosphere rarely feels acute distress; when the oxygen in his blood is about half normal, he faints quietly and his brain may show a few minor electrical disturbances, but often none. Yet, if recovery occurs, the period of oxygen lack may leave traces for many days or for life.

This is the Achilles' heel of the brain. Its need of oxygen and sugar is perpetual and exacting. Lack of sugar it can signal, but not lack of oxygen. The agony of suffocation, the gasping and craze for air, is in fact due to the accumulation of carbon dioxide in the blood from the burning of the body fuel; and this the brain does feel, choked as it were by its own smoke, even though it cannot detect when the dampers are closing down.

No one has yet succeeded in relating the electrical features of the living brain with the now classical but still disputed categories of mental disorder—schizophrenia, manic-depressive psychosis—or with the symptom groups: paranoia, delusion, hallucination, obsession, compulsion, depression, agitation and the like. This has been a sad disappointment, but we may be comforted to recall the unexpected complexity of the analysis of the simplest form of learning. The electrical signs of mental disturbance could well be a minor aberration in one or other of the operations leading to the formation of ideas. We should not detect such differences in a passive patient. Studies of the brain responses to various types of stimulation in cases of mental disorder are now in progress in many laboratories, and first results suggest that such experiments will clarify much of the inherent weaknesses and self-restoring powers of individual brains. The difficulty is that the more refined the method of study, the more striking is the individuality of each subject and of each instant in the subject's life. We are still lost in admiration of the cerebral universe and have scarcely begun to name the constellations and trace the course of individual planets in the electrical firmament.

Willingness to ascribe mental differences and disorders to brain characters and diseases varies from generation to generation. There are physicians now living who remember the time when general paralysis of the insane was the commonest form of serious mental disorder, and was ascribed to the effects of excessive travelling, since it was most common in sailors, salesmen and locomotive drivers. When it was shown to be due to the spirochaete of syphilis, prevention and cure became a matter of pharmacology and common sense. At the present time, many disorders that seem essentially mental are being attacked by physical means; there is less prejudice against this today than there has been for several hundred years.

Physical interference with personality by operating on the brain has been accepted with quite extraordinary public equanimity. Psychosurgery has developed through bold and reasonably safe operations by which parts of the brain are removed, destroyed or isolated, in patients with only mental symptoms. Egas Moniz of Lisbon, who first demonstrated the practical value of such operations, recently received a Nobel Prize; his pre-frontal leucotomy, the cutting of the nerve fibres that connect the front of the brain with the rest, has been performed on thousands of persons. Very few deaths have been reported, and some astonishing "cures" have been achieved, even in patients who had been mad as hatters for years. The most promising results appear to be more or less enduring changes of personality. People with silly fixed ideas which stop them doing anything useful sometimes say after the operation that they still have their obsessions or delusions but do not worry about them any more. Even when the main trouble is intolerable pain—due to whatever cause —leucotomy seems to help the patient not to care so much, though the physical cause of the pain is still there. Many leucotomised people have returned to a full and happy life and indeed are often particularly agreeable and easy to get on with. An American surgeon has said that the nicest people in his hospital are the ones with the little scars where the horns used to grow.

This is not a refined manner of treating the brain. The operation is sometimes effected through the eye-socket and needs no special tools; it could have been done by a neolithic savage equipped with nothing more than his flint scraper and the conviction that mental diseases can be attacked by physical methods. Some ancient skulls do in fact show trephine holes where there is no evidence of organic disease. In the present trend of thought, until the contrary is proved, it will be assumed that mental processes have physical representation in the brain in some form, and when the mind is disturbed surgical methods will be used where others fail.

Operations such as leucotomy, however, are mainly of symptomatic benefit to the individual; their long-term biological effects on society are hard to assess. It must be remembered that if the symptoms only are relieved, in a disease which is not understood and which may have hereditary factors, the effect may be ultimately to increase the prevalence of the disease. For thousands of years the insane and eccentric have been discouraged from breeding by Church and State; so neither the disapproval of the operation expressed by the Vatican nor the decree of the Kremlin forbidding it is surprising. Without considering what part heredity plays in mental disorder, leucotomised mental patients are put back into circulation and, as noted, they can be very pleasant and easy-going. Thus, courageous and inspired surgical invention can raise quite serious problems of the relation between individual welfare and the future of the species.

The effect of leucotomy on the electrical rhythms of the brain is variable and correlates little with the clinical results. Usually there is some increase in alpha activity after the operation, and delta activity may be prominent immediately after it but disappears within a matter of months. What have been described as the personal characteristics in the behaviour of alpha and theta rhythms are not greatly altered by leucotomy, or by electric shock therapy.

In other respects electric shock therapy also alters the personality to the extent of allaying some of the most trying symptoms of mental disorders and psychological distress. It has quite surprisingly won public confidence—one might almost say popularity. But again, though a less drastic procedure than leucotomy, it must be confessed that very little is yet known about its influence on the life history of a brain, beyond what has been empirically demonstrated and clinically observed. When first introduced it was hoped that it would throw some light on epilepsy, to which its convulsive effect is related, but beyond the confirmation of certain therapeutic aspects of epilepsy, mentioned elsewhere, it has not yet brought any major revelation such as those obtained by non-clinical techniques. It is essential, however, that research should continue to follow this and other experiences in psychosurgery.

An axiom demonstrated by the heroic physical treatment of mental disorder is one that has been emphasised in an earlier chapter: only a few brain functions are permanently located in one part of the living structure. A person who has suffered an extensive brain injury, or submitted to prefrontal leucotomy, usually displays some changes in personality; but re-education and discipline can often re-establish a great part of the missing pattern. In the early days of psychosurgery the patients were frequently left to recover without psychiatric or psychological assistance; it is now realised that the good effects of physical intervention can be greatly amplified, and the bad effects diminished, by carefully planned and rigorously maintained personal and social re-education. Even a mutilated brain retains plasticity and resilience. Although it is less flexible than before surgery, it can still learn, still adapt to environmental stress.

This principle applies in a less welcome fashion to the tendency to relapse after physical treatment. A patient with a characteristic personal symptom complex may obtain great relief from a leucotomy operation, may worry less, return to work, fit in better with the social pattern; then, perhaps, ten years later, the symptoms may return in precisely their original form. We know that in the brain there is no possibility of physical healing—a nerve cell is never replaced. A brain can no more grow a new frontal lobe or re-establish a severed pathway than the trunk can replace an amputated limb. Recovery of normal function and reversion to pathological disorder are both examples of the principles of tireless search, of parsimony and of plasticity, discussed earlier in connexion with some simple models.

One aspect of widespread submission to psychosurgery the research worker must welcome unreservedly. The present abundance of subjects with carefully planned, restricted and reasonably uniform brain injuries presents a valuable opportunity to study the re-establishment of functional patterns; otherwise the casualties of road and battle were his only material.

The cerebral circumvention of crippling obstacles, the resumption of even pathological behaviour patterns, is a phenomenon of such a personal nature that it should be common ground for students of mind and brain alike. To the latter it already suggests that when the physiological count of the content and structure of personality differences is made, it will be as detailed and important as the description given of them when psychic factors were deemed sufficient to define psychiatric states. Pure and applied psychopathology already has an enormous literature and a powerful tradition; but the material evidence from this field, from the physiological standpoint, is hard to assess, so serious is the conflict of testimony. As yet there is very little common experience on this common ground.

In the realm of normal psychology, the difficulty is just as great, where so many currents are turbid with sectarian strife. Several schools of psychology have set up various schemes of types with which correlation can be sought; some are based on physique, some on results of personal enquiry, some on tendencies to pathological extremes; and the typology devised by the Pavlovians has been outlined here. It would be gratifying if EEG studies could be made to relate to any of these classifications; but so far no precise or even suggestive correspondence has been established with any of them.

For the time being, then, we must be satisfied to summarise the material intimations of personality found in the electrical activities of the living brain, without academic psychological correlation. They have at least the merit of being not opinions or elements of a theory but facts recorded in experiments which can be repeated. Brainprints, the records of the electrical rhythms, beyond giving personal identification such as that of fingerprints, can be obtained even before the birth of a child, and from then onward they display in various frequencies and amplitudes the maturing characteristics of mental development. This is associated, though not in a simple manner, with the maturing anatomy of the brain. The diversity is so great as to seem random; but refined automatic analysis of the main rhythms reveals in all of them certain features which can be correlated with the mental experiences of the subjects, providing the data for a very diversified classification of types of all ages. Childish behaviour in adults was shown to be related with characteristic rhythms of childhood, and self-control demonstrably measurable. What appears to be a scanning for pleasure was found dominant in childhood until, as the brain matures, the adult scanning of the alpha rhythms takes its place. Study of the alpha rhythms themselves revealed the personal characteristics of three different, and sometimes incompatible, ways of thinking. There was a hint that, in connexion with the unexpected complexity discovered in the simplest form of learning, certain signs of mental disturbance could be related to minor aberrations in one or other of the seven operations. Study of the alterations of personality effected by the physical treatment of mental disorders confirmed some of the principles of organic construction exemplified in working models. The popularity of psychosurgery suggested some cautionary reflections while providing the brain physiologist with an invaluable field of research.

Finally, it must be recalled that these are only the first fruits of this sapling. As recently as 1946 an eminent physiologist could write: "It remains sadly true that most of our present understanding of mind would remain as valid and useful if, for all we knew, the cranium were stuffed with cotton wadding." Ten years ago, in fact, there was no precise knowledge about any of the matters discussed in this chapter.

CHAPTER 9

Beyond the Waking Scene

We sleep, but the loom of life never stops and the pattern which was weaving when the sun went down is weaving when it comes up tomorrow.

Henry Ward Beecher (1813–1887)

THE PICTURE of the living brain outlined in earlier chapters was one of incessant, often fruitless, activity. Even when physiological or mental performance seems at the lowest ebb, electrical rhythms and surges are sweeping through the brain fabric in a shimmering procession; and, even in the pursuit of meaning, the greater proportion of the brain's wagers, its prospecting, must be vain. Yet, nevertheless, there is no respite for the electro-chemical generators, whose only relaxation is to pass from individual preoccupation to group callisthenics.

When we close our eyes our alpha rhythms strengthen and join up into long trains of sinuous pattern; this may rest our eyes, but as we draw the blinds the lights within seem to be turned up to greater brilliance.

In fact the increase in alpha activity during rest does not, we suggested, imply a fundamental paradox, but rather is a sign of unsuccessful search for pattern. The brain is unique in being constantly speculative and expectant, and consequently often frustrated and disappointed. We can use such terms legitimately because we recognise, in simple laboratory ex-

periments or in crude working models, conditions to which
they apply in all their senses. Some other organs can relax in
hours of ease. The body muscles maintain only a skeleton staff
on shift work when the limbs have no job to do; a few fibres
contract from time to time so that some tone is preserved. The
heart muscles, on the other hand, have a full-time occupation, but their lack of leisure is compensated by the relative
simplicity of their function; only in moments of stress is the
rhythm of their activity modified.

The perpetual vigilance and adaptability of the brain depends, as we have seen, upon its emancipation from menial
tasks, upon delegation of the bodily chores of homeostasis to
primitive centres, the original and subordinate peasantry, as
it were, of the central nervous population. Viewed imperson-
ally from outside, we might have expected that such a system,
being insulated from all but major emergencies, would be
both stable and indefatigable. And considering the intricacy
of its functional organisation, the stability of the brain is remarkable. But tireless it certainly is not.

Perhaps the most obvious sign that there are degrees of
vigilance in cerebral attention is the power to close the eyes.
Even the frequency of blinking has its significance; in some
neurotics and in certain types of insanity the rate may be ten
times higher than in normal people—not to be confused with
the repeated blinking of some people when they are deep in
thought.

Only the visual system can be cut off in this way from sensation. Experiment has shown that lowering the eyelids is more
than the mere closing of a shutter. As the lids fall, the eyeballs
roll upward, so that the cornea is wiped and protected; at the

same time, certain limiting or protective mechanisms within the brain are suspended for an instant. This can be seen most clearly by application of the flicker technique. In many subjects in whom the response to flicker with eyes open or closed is normal and restrained, if the flickering stimulus is applied just at the moment when the eyes close, a grotesque rhythmic discharge appears in almost all brain areas, with an appearance and accompanying sensations very like a brief minor epileptic attack. For this shock effect to be seen it is necessary for the subject to want to close his eyes in the light and to succeed in doing so. It does not happen if, instead of closing the eyes, the room light is turned off as the flicker is turned on; nor if the eyelids are closed forcibly; nor if the eyes are closed in the dark; nor if the subject wants to close his eyes but is prevented from doing so. It seems that, with the decision to close the eyes, goes a relaxation of some safeguard against over-activity.

In other people, including many who are liable to give exaggerated responses to flicker, small irregular sharp waves appear from the back of the brain while the eyes are open and gazing without extreme attention at a patterned scene. These waves disappear when the gaze is fixed on a point and when the eyes are shut. Christopher Evans, who first drew attention to them, has called them lambda waves. Rather similar effects are seen when a spontaneous discharge which is being held in check by an inhibitory process is partly released by a drop in intensity of the inhibitory influence, or when rhythmic stimulation is used to invert the action of an inhibitory mechanism. In this latter case the stimuli are inhibitory, but at the end of each of them an excitatory rebound occurs and a paradoxical excitation is seen, as though the system were responding to each stimulus, whereas actually it is responding to the *end* of each stimulus.

Closing the eyes, then, involves some sort of momentary dis-inhibition, which promotes abnormally wide dissemination of any signal that may happen to enter the visual pathways at that moment. It is as though the sentries were dismissed just as the drawbridge is raised; if they leave their post a little before the bridge comes up, then, just for that moment, an assailant can penetrate an otherwise impregnable keep.

The repose we get by shutting our eyes is not, of course, enough to refresh more than a small part of the brain; in doing so, many people—the M-alpha types—merely pass from a public to a private picture show. But closing the eyes is an essential preliminary to going to sleep, a process which can be followed quite easily with the EEG. (See Figure 20.)

When a subject begins to feel drowsy, the first sign is usually a *reduction* in alpha rhythms and their gradual replacement by something more like the theta activity, mainly at the back and sides of the brain but spreading occasionally to all regions. At this stage the subject is easily roused and not yet really dozing; it is the beginning of that delicious state when consciousness is consciously waning. If the subject is trying *not* to shut the eyes, he will find himself seeing double; if reading, he will go over a paragraph time and again without getting its meaning. Some have called this state "floating," a metaphor that well describes the freedom from bodily care which it promotes. This is the clue to understanding the nature of this stage of torpor; it is as though the whole body were being shut out from mental view, as the outside world

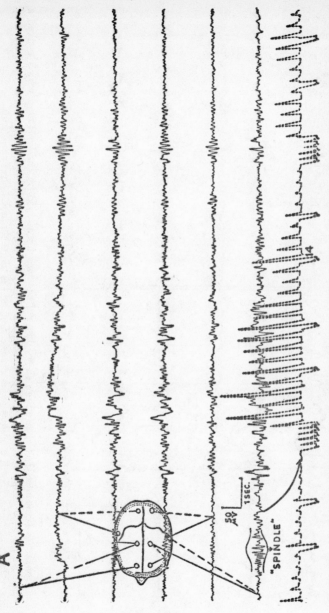

Figure 20. ". . . these rhythms are the wardens of brain function." Three Stages of Sleep Recorded. (a) Light sleep—"floating." Half-waking dreams may be associated with the disturbances shown in the record.

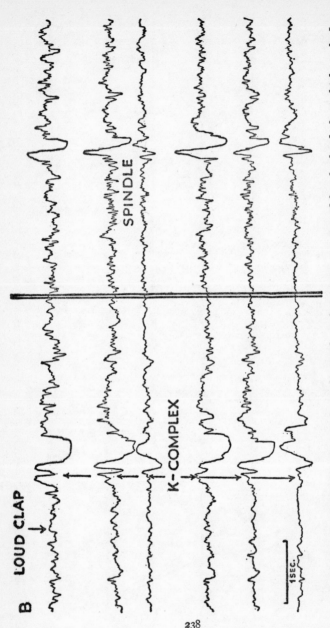

Figure 20 (continued). (b) Medium sleep. The subject was not roused by the loud clap which evoked a response possibly associated with a protective dream. Later, a "spindle" followed by a similar complex may have been associated with a spontaneous dream. Calling his name immediately roused him.

238

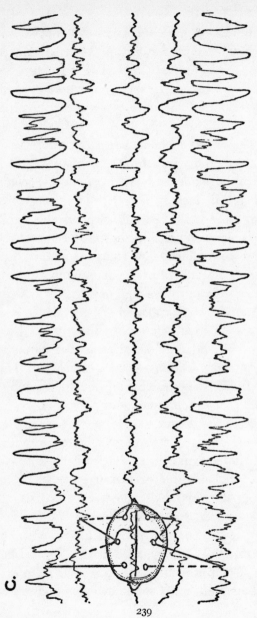

Figure 20 (continued). (c) Deep sleep. When the slow waves, best seen in top and bottom lines, are present, the subject is hard to rouse.

is shut out by closing the eyes. Here of course there is no mechanical or anatomical shutter for the muscles and joints that can deceive the brain into neglect of its supervisory commission; there is however an elaborate mechanism in the base and stem of the brain that, when fatigue or custom dictates, inexorably weakens the significance of the incoming flood of sense data. This is the moment when the exhausted driver begins to persuade himself that the road is so straight that he can safely drive a while with his eyes shut—a comforting thought though it be his last.

At this stage the brain can still respond electrically and functionally to incoming signals, but the electrical response begins to show a prominent slow component, and the spread of the faster excitatory effects is more limited and more transient. This is the objective sign of what we called above a *weakening of the significance* of the signals. It is not the direct transmission of the nerve impulses to the projection areas that is inhibited, but rather their dissemination—the first two operations of learning—that is attenuated. The reader's eyes can follow the lines of print but the meaning of the words escapes him and, merging into private fancy, his bedtime story becomes a dream.

The electrical changes in this first stage of drowsiness are clear enough to be exploited as a selective warning device for those who are liable to doze in dangerous situations. The resemblance of the rhythms to the theta rhythms of anger and frustration means that such a device would also flash a light, or ring a bell, or nudge the subject, when he was losing his temper. Worn by hard-driving motorists, theta warning-sets would probably save more lives than do motor horns; and they might assist self-knowledge and self-control.

If we follow a drowsy subject further along the path to oblivion, we soon see other changes; slow irregular waves appear, and with them, from time to time, bursts of smaller, faster spindly waves with a frequency of about 14 c/s. The subject is now asleep and if we waken him suddenly will start and look confused; he may deny that he was sleeping, or he may say that he has just had a dream.

There is some evidence that the 14 c/s spindle rhythms are associated with dreams, but this is hard to prove. Most dreams are very short, compared with the time that seems to elapse during them. Very often outside stimuli are woven into dreams in such a fashion as to diminish the urgency or importance of the stimulus. Such dreams help to maintain sleep. Experiments with controlled stimuli show that a sleeper may turn a powerful sound stimulus lasting, say, ten seconds into a continuous feature of an elaborate dream-sequence which seems to last for days. The sense of touch is not always so amenable. Prompted by some momentary sexual stimulus, the sleeping brain, master of fiction, will weave a romance around a heroine of its own invention, a post-bedtime story that sometimes ends in provoking the stimulus it stilled.

As slumber deepens, its characteristic patterns become more pronounced and the effect of stimuli assumes a new appearance in the EEG. While the slow delta waves dominate the undisturbed pattern, any loud sound or firm touch evokes a complex electrical discharge from all areas of the brain. This discharge was called the "K-complex" by Loomis and his colleagues. It resembles considerably the pattern found in certain types of epileptic patient during waking, but in these latter it is then often associated with a brief convulsive spasm of the limbs. In the sleeping normal subject the appearance of a K-complex is a sign that the sleep is deep and hard to disturb, and it seems to act so as to muffle the arousal stimulus. (See Figure 20.)

The most prominent feature of the K-complexes is the train of large slow waves which reach every region. If such a figure occurred in the waking state we should regard it as a sign of profound pathology. But in sleep they have a beneficial role. Like the dream that explains away the unwelcome alarm bell, K-complexes seem to flatten the significance of intrusive signals, so that the sleeper's standard of meaning is stretched beyond the tolerance of waking life. An exclamation that would evoke an immediate response in a waking subject must be repeated a dozen times at irregular intervals to arouse a sleeper, and at each repetition will appear the faithful watchdog at the Gate of Horn, admitting to consciousness only those dreams that stem from hard reality.

Sleep, in a restive hostile world, is a hard stake to win; in jungle or tenement there are many voices that cry out against it, and of these all but one or two are vain. Only the infant's wail for the mother, the call to work for the breadwinner, the tapping on the window or the lover's murmured plea, faint though they be, retain their insistent meaning. The rest—the chatter of the tribe, the creak of rigging, the drumming of wheels on rails—are lullabies, not signals.

Rhythmic stimuli, we know from our cradle, are sleepmakers; and where pattern lulls, its interruption rouses. Who, on a long night journey in a sleeping car, has not found himself awake to silence or the gentle hiss of steam in some wayside halt, and wondered that the racket of the train *in ceasing* had aroused him? This too we can see in the wave patterns of the sleeping brain; a steady sound or rocking movement leaves no trace on the constant delta waves; but when it ends, at first a K-complex appears, and if the steady sound starts up again the pattern reverts to that of slumber. But a longer interval evokes no further K-complexes; instead, within a few seconds, the pattern shifts up through the stages we have described, and abruptly the sleeper wakes.

As in all behaviour of the waking brain, the hues of personality gleam through these shadows. The length of each stage of sleep and the pattern of awakening, the form and extent, the abundance and efficacy of K-complexes, the frequency of the spindle-waves of dozing—all these features are assembled in the nocturnal rhythms with as personal a character as the waking ones. There is some evidence that the "type" of waking record is associated with the type of sleeping pattern, but the statistical significance of these observations is not certain. There is still a great deal to learn about the physiology of sleep and its relation to brain functions as a whole. The patient studies of the Gibbses in America have already cleared up many outstanding difficulties. The experiments are tedious and time-consuming beyond the average, even for this sort of work. Even to write about sleep induces it!

There is no clear explicit reason why the brain should need to spend a third of its life in repose. Is it for the sake of some other organs that the master centre closes down its receiving and transmitting channels? Is there some subtle chemical by-product of the brain's fierce sugar-furnace that accumulates in toxic quantities during sixteen hours of waking work? Or is our precious sleep but the genetic trace of the futility of nocturnal adventure in a wilderness of wide-eyed cats and jackals, an heirloom from remote forbears who were perhaps better equipped for the fashioning of well-hidden nests than for the duel in the dark?

Perhaps this apparently redundant rhythm arose even further back in evolutionary history, when the first vertebrates emerged from the incubator sea and felt, for the first time, the chill of evening. Such creatures, equipped with elaborate nervous systems, air-breathing lungs, and five-fingered limbs, could escape the material tyranny of the ocean only to find themselves at the mercy of the seasons. The temperature of the sea varies but little, scarcely appreciably from day to night, and only by a few degrees from summer to winter. Sea creatures can sustain an even life though the temperature of their bodies follows the warming and cooling of their liquid home. But on land the conditions are too extreme to maintain the tranquillity which a nervous system must have. If the temperature falls by 20 Fahrenheit degrees, nerve impulses will travel at half speed, and the fine pattern of nervous function will lose its texture. In the first chapter we allowed ourselves to conjecture on the way of life of these early explorers between the tidemarks and on the shore. Reptile or amphibian, as they established their dominion over dry land in the many million years of summer which was the carboniferous age, they must have suffered at noonday a delirium of over-action, and as night fell a sluggish torpor. Fortunate and fertile were those beasts who submitted with grace and discretion to the inevitable rhythm, curling away quietly in a private cranny as the impulses lagged in their cooling nerves.

In the private sea of our bloodstream we carry a chemical miniature of the primeval sea. So perhaps in our brain we inherit the need to retire from active strife as darkness falls, even though with our hardwon homeostasis and technical

ingenuity, our fur coats and our electric blankets, we can
maintain our body temperature within a fraction of a degree.
Even now, as many a brave man has testified in the burning
cold of the poles or mountain peaks, when all these defences
fall and with them the body temperature, it is not a struggle
or a convulsion that leads to death, but a peaceful sleep. Carrying the conjecture a step further, we discover in the slowing of the brain rhythms a trace of that nervous deceleration
that lulled our predecessors.

In most complex machines, when some part is fatigued or
ill-adjusted, it is better that the whole mechanism should stop
than that its function should continue with the possibility of
disaster. An engineer describes an arrangement that ensures
an orderly shut-down when a part gives way as "failure to
safety." In some dangerous machines such mechanisms are
given the more dramatic name of "dead man's handle," meaning that, should the driver lose control, the vehicle or turret
will come to a standstill rather than career away without a
guide. In a mechanism so intricate and so delicately poised
as the living brain we should expect to find, if not a dead
man's handle, at least a sleeping man's knob; and it would not
be altogether surprising if, with our genetic history, the knob
were turned by the inherited tradition of our bodily structure.

As long ago as 1938 I suggested that delta rhythms "represent a true change in the natural electrical period of the
cortical neurones, a change which is usually ominous but yet
reversible if the cause is removed. During delta activity no
useful work can be done by the neurones concerned. Sometimes the delta waves are so large that we may suspect them
of paralysing the cortex by electrocution, as it were, and we
may speculate as to whether this may not be their special

function in certain conditions, just as the function of pain is sometimes to immobilise an injured part. In order to explain the occasional great size of the slow waves it may be necessary to invoke a series connection for the cortical cells, similar to that evolved from muscle cells in electric fish."

As we have already seen, slow rhythms are found in infants, and in adults with organic disease or injury of the brain. Slow components are prominent, too, in epileptics between seizures and during attacks when there are no convulsions but only loss of consciousness. All these conditions have in common the need for protecting the brain from the consequences of its own complexity. The serious results of loss of control and protection are seen when an epileptic has a major convulsion. Then, very rapid electrical discharges predominate, and the whole system is thrown into revolutionary chaos. Toward the end of the seizure, slow waves again appear and the rapid convulsive discharges emerge only in the troughs of the slow waves. Such considerations have suggested a phylactic, or protective, hypothesis to account for the slow electrical rhythms, according to which these rhythms are the wardens of brain function, limiting the consequences of excessive or ill-co-ordinated activity. Seen in this light, the K-complexes of sleep take their place with the slow waves of infancy, epilepsy and other conditions of incomplete control, as censors of the news from foreign parts, ensuring apathy if they cannot maintain discipline.

Facile parallels are sometimes drawn between these electrical studies and speculations of a more philosophic trend, such as the concepts of the psychoanalytic school. There is much to deride, and not a little we may be uneasy about, in that doctrine, but to give Freud his due, the founder of psy-

fatigue and ennui? We may be tired without being sleepy, bored without being tired.

Fatigue has many forms, but few of us have ever reached the limit of our endurance. When we have done hard muscular work, we feel that our limbs are tired, and we limp or slouch like cripples. But the fatigue is not likely to be in the muscles—an extra incentive will evoke an extended response. Robert Schwab in Boston has shown that if a man has to hang from a bar by his hands "as long as he can," he may drop after a minute; if he is urged to better someone else's record, he may double his time; but if he is offered ten dollars he may hang on five times as long. From an active group of muscles the brain receives signals reporting the effort they are making and the extent to which this matches the effect intended in the action. After some while, pain is felt; the active elements file a routine complaint of ill usage some long while before their efficiency is actually threatened. To such reports, as to all others it receives, the brain can assign the degree of significance that best fits the whole situation. Persuasion renewed, danger, training, pre-occupation, experience, hysteria, drugs—all can modify over a very wide range the meaning of fatigue and its effect upon the whole organism.

In the EEG very little trace is left by fatigue of ordinary dimensions, but the weariness that follows 24 hours' work without rest does result in some change in the electrical patterns of response. The resting records show more readily the preliminary changes of floating and drowsiness, as one would expect; but the responses to stimulation are often quite dramatically changed, particularly if the subject still feels bound to keep awake. One subject, whose normal re-

sponse to flicker was slight, and accompanied only by negligible visual illusions, after a day and night of hard anxious
toil exhibited an enormously exaggerated flicker response extending far into the temporal and frontal lobes. At the same
time he experienced for the first time in his life a vivid visual
hallucination: "a procession of little men with their hats pulled
down over their eyes, marching diagonally across the field."
It is easy to imagine that the little men, tired, leaving work
and half asleep, were symbols of the subject's exhaustion,
easily evoked by the flicker, while the guards against such
fantasy were overpowered by stern necessity.

The brain can register fatigue and usually surrenders to it
long before the other organs have called on their reserves; but
we should consider it as being, like sleep, more a positive
turning-off than a negative running-down. The feeling of
fatigue can be accelerated and amplified by boredom, by
a monotony of insignificant signals; it can be postponed or
diminished by drugs such as caffeine and amphetamine, both
of which have slight but measurable effects upon the normal
electrical rhythms of the brain.

Hypnosis, too, can influence the onset of fatigue. And here
we come on a strange anomaly. The hypnotic trance is a distinctive state; a hypnotised subject can perform easily feats
which he would otherwise find difficult, can support pain
which he would otherwise find intolerable, and recall incidents he had forgotten. In light anaesthesia, too, such changes
are seen. But whereas an anaesthetic produces regular and
striking changes in the brain rhythms and responses, hypnosis
has little or no effect on them. In most experiments there has
been almost no significant change in the EEG while the subject is hypnotised, even when the trance is quite deep and

behaviour apparently much changed. When a hypnotised subject with his eyes closed is told that they are opened when they are not, his alpha rhythms continue just as they do in normal conditions, although he seems to believe what he is told. Conversely, they do not appear if he is told that his eyes are closed when they are not. Yet his declared impressions follow the hypnotist's suggestions with great fidelity.

Hypnosis shows none of the electrical features of natural sleep; indeed, the more carefully we consider the subject's state, the less it seems to resemble anything we know of sleep. Awareness is not lost, but heightened—restricted, it is true, to specific categories of stimuli, usually the hypnotist's voice and suggestions. The significance of events is not reduced, but absurdly emphasised—the power to learn is extended from its proper field of significant pattern to any trifle the hypnotist may fancy. So exaggerated may become the extent of suggestion, that bodily reactions of a serious nature—blisters, bleeding, cramps or swellings—may follow on injudicious suggestions of injury or passion.

The power of conditioning, as demonstrated by laboratory experiments, the cult of Yoga, and the ailments of hysterical origin, seem to share the same mechanisms. In hypnosis, again, we see how wide and deep is the dominion of the brain over all other organs and functions. But in this state the rules of conditioning seem to be waived. There is no reward or punishment, no dignified succession of neutral and specific stimuli. In some way the hypnotist gains access to the inner workings of the learning mechanism without diverting or distorting the basic properties of brain function, so that all the pathways of association and stores of experience are intact though their contact with the outer world be limited to his chosen vehicle. It is well known that some people are far more easily hypnotised than others and that, with very few exceptions, although the hypnosis may break the rules of learning, it does not transcend the boundaries of habit or principle. Few hypnotised subjects can be made to carry out suggestions which are indecent or dangerous according to their standards of conduct or safety.

Nobody has yet offered a plausible complete explanation of the hypnotic state, but an evident invitation to investigation lies in that readiness to assume significance which has been emphasised as the dominant attribute of brain function. The device by which the hypnotist contrives the plasticity of dreams without the apathy of sleep is a matter for further study and experiment.

To many people the evidence of the power of brain over body is so impressive that it seems to extend itself quite naturally to another category altogether, the influence of mind over matter, and to suggest an indefinite merging of such physiological curiosities as we have outlined into some transcendental realm of spiritual experience. We must confess at this stage that no study of brain activity has thrown any light on the peculiar forms of behaviour known variously as secondsight, clairvoyance, telepathy, extra-sensory perception and psychokinesis. It has often been suggested by those seeking a material basis for otherwise unaccountable behaviour that the electrical activity of the brain might be the mechanism whereby information could be transmitted from brain to brain, and that the electrical sensitivity of the brain might be a means of communicating with some all-pervading influence. Quite apart from any philosophic objection there may be to such an argument, the actual scale and properties of the brain's electrical mechanisms offer no support for it. The size of the electrical disturbances which the brain creates are extremely small. In fact, they are about the size, within the brain itself, of a received signal which is just intelligible on an average radio set. More crucial even than this, their dominant frequencies are far below the range of radio channels, below even the scale of audible frequencies. At ten cycles per second, the average frequency of the alpha rhythms, any electromagnetic signal transmitted through space would have a wave length of thirty million metres.

The familiarity of radio signalling around the world has popularised the notion that any signal once generated may be propagated indefinitely through the chasms of space, so that all events have an eternal quality in some attenuated but identifiable form. This is not even approximately true; for any signal, however propagated, weakens with its passage until its size falls below the level of noise and interference in some locality. Beyond this point it can never be detected, however great the resolution and selectivity of the receiver. If we consider the largest rhythms of the brain as casual radio signals, we can calculate that they would fall below noise level within a few millimetres from the surface of the head.

Even if we ignore these physical characteristics, the observations reported on extra-sensory phenomena seem to exclude any such approach; for there is no evidence that screening of the subject, or distance between sender and receiver, has any influence on the nature or abundance of the effects described. Furthermore, it seems to be one of the cardinal claims of workers in this field that a signal may be received before it is transmitted. If we accept these observations for what they are said to be, we cannot fit them into the physical laws of the universe as we define them today. We may reject the claims of transcendental communication on the grounds of experimental error or statistical fallacy, or we may withhold judgment, or we may accept them gladly as evidence of spiritual life; but it does not seem easy to explain them in terms of biological mechanism.

CHAPTER 10

The Brain Tomorrow

I should be glad if, when people come to a clear understanding in natural science, they would stick to the truth, and not go transcendent again after all has been done in the region of the comprehensible.

Goethe, *Conversations*

THE SALIENT physiological facts about the living brain known at this time of writing have now been told in more or less detail as their general interest seemed to require. There remain one or two things to be said about the future, things which might still be inoffensive to the sage, who did not exclude scientific prediction from the truth. But first some of the matter about which, it is hoped, we have come to a clear understanding, may be summarised.

A sketch of the evolution of the nervous system led us to the emergence of the human brain and its emancipation from the menial bodily tasks. Next was traced man's tardy recognition of the organ that alone sets him apart from other living things, the slow rise of interest in it, the rapid development that took place in electroencephalography as soon as it was realised that the patterns of electrical activity which the new technique revealed were not random effects but vital signals, and might even have among them the physical counterparts of mental events. In order to see the significance of these patterns, some elementary facts about brain functions in relation to the sense organs were given, and some recently devised techniques were described by which fresh knowledge has been obtained about such diverse matters as epilepsy, hallucinations and the scanning functions of the electrical rhythms. The construction of working models to imitate life, not in appearance but in behaviour, was discussed, and they were found to be not only imitative but informative.

Two main paths of research were then traced, stemming respectively from Berger's discovery of the electrical rhythms of the brain and Pavlov's demonstration of the measurable conditioning of sensorimotor reflexes; and some little-known aspects of the latter's work were described. These two paths converged in the recognition of a logical affinity between certain electrical responses and the process of learning—on the one hand a unique mechanism, on the other a unique function, of the human brain.

After considering the scanning theory which this coincidence suggested, the climax of recent research was reached in an analytical discussion of the learning process. Laboratory observations and experiments had made it seem necessary to postulate a minimum of seven operations in the mechanism of learning. A working model was described in which the performance of these seven operations had been provided for. Here again, even at this high level of reflex complexity, the model not only satisfied all the theoretical requirements predicated but discovered phenomena for which no arrangement had been made—phenomena which nevertheless were found to correspond to those of the living brain under similar experimental conditions.

In traditional physiology the equivalence of similar organs is implicit; apart from matters of taste, all idiosyncrasies are pathological. In the brain, one is tempted to say, the reverse is true, or at least more true; idiosyncrasy is the rule and equality a myth. The brain is essentially the organ of personality. So brain physiology necessarily takes note of the variations normally found from brain to brain; and it was shown that already an entirely new stock of data had been accumulated concerning measurable differences of personality. These intimations were found to be already plentiful and positive enough to suggest the possibilities of a scheme for grouping characters at various ages, not in correspondence with any of the divers groupings of professional or academic psychology, with which indeed no correlation has yet been found, but on its own basis of measurable results in experiments that can be tested by repetition and cannot be thwarted by coaching.

Some thoughts about the future of the brain may have occurred to the reader at this point. There is one obvious aspect of the matter. When some of the new facts about the development of character become generally known—for instance, that tendencies to violence can be detected, that self-indulgence and self-control can be measured, together with the effects of encouraging or discouraging them—the brain specialist, be he psychiatrist or physiologist, will be bound to concern himself with its future treatment, particularly in education. But what of other implications?

After bowing to Goethe's sage predilection, and after what has been said in the preceding chapter about transcendental phenomena, we must not let prediction merge into prophecy. The difference is of course fundamental. Prediction is an extreme of reason, prophecy an extreme of emotion. Extremes meet only when the mind is clouded; at other times we know whether the brain is fulfilling its own essential function of prediction or is serving the elemental cause of the emotions.

The evolution of the human brain itself, and the course of our knowledge of it, converge momentously today, reacting on each other. The brain is not what it is by virtue of any informed planning; its education has been evolved without regard for its functional constitution, about which nothing relevant was known ten years ago. This cannot go on. For teachers and educationists throughout the ages the brain has been a Black Box; its working has been known only by comparing output with input. The brain came to maturity in complete ignorance of its own existence; its training has been empirical and inferential.

When tracing the processes of learning in a previous chapter, a point was reached where it could be said that the critical operation was the recognition of pattern, an abstraction. This was advanced without thought of any parallel in the evolution of the brain itself; but there is in fact a very striking one, an abstraction at a critical phase which may be regarded as a functional recapitulation of the organ's growth. Freed by homeostasis for intellectual pursuits, the supreme abstraction of the brain was indeed the mind. And *mens sana* until recently has been the only watchword. But changes are taking place in which the material success of psychosurgery is significant. From the confusion of metaphysics and psychoanalysis, abstractions of an abstraction, the thinking brain has turned eagerly to the first possible glimpses of itself. The millennial period of its unconscious evolution ends before the mirror; a new phase begins.

The educationist is of course primarily interested in this convergent crisis, of which the crisis in education itself is a symptom. When we consider what has been achieved in that field, the mass improvement and the peaks of individual achievement, the method of the Black Box must be revered; and many of course will want it to be retained; a Black Box and a familiar abstraction make fewer demands on one's time and attention than do a living brain and its unfamiliar functions. Also there is some apprehension, as among the spiritual-minded in the days of anatomical ambition to locate the dwelling-place of the mind, that one might look into the Black Box and find it empty. For those who think the learning process is incomplete because an idea emerges from an abstraction, "mind" is a necessary but is not an ultimate conception; intellectual materialists, for them the spirit must always be made flesh; of all miracles, transubstantiation is the most comforting. Or they may "rationalise." As Crawshay-Williams put it in "The Comforts of Unreason," the word "mind, though originally a label for a complex of mental events, often becomes hypostatised into an entity" and, he adds, referring to Eddington's disclaimer of the larger entities of orthodox religion, "once an immaterial entity is born it is surely futile to excuse it by saying that it is only a little one."

We are not going out of our way to nurse this indiscretion; psychological baby-sitters are legion. Any discussion of mind except as a function—the supreme function—of brain, lies beyond the scope of this work and must always remain outside the purview of physiology. The physiologist, viewing in his modest workshop the inexplicable electric tides that sweep through the living brain, knows that the bobbing of his float must mean some Leviathan is yet uncaught; some great idea nibbles his bait and slides darkly behind the laughing waves.

But he would be happier not to dub it Mind; he would prefer to call the one that got away—Mentality, thinking of it only as a relation of dimensions, in the same class as Velocity. We speak often of the "craze for speed," but only a truly mad mechanic would peer into his engine for the component of its velocity. So, even in a fever of interest in mental problems, no sane physiologist would look for a mechanism identifiable as Mind; but he may quite reasonably say: "At one time behaviour was *so;* later it was *thus*—the transformation of one mode into the other I will call *Mentality.*" In the chapter on models of behaviour we saw how even in the very simplest system, with two active elements, multiple interconnection between elements gave several modes for which simple observation was useless. The study of mentality, product of interlaced patterns far too numerous even to write down, cannot therefore be defined, let alone solved, by contemplation of behaviour, however patient the experimenter, however sensitive his instruments.

Nor is the proposition made any clearer or more tractable by glib assertions about studying the organism "as a whole." In revulsion from the cult of dissection, of particulate experiments on isolated bits of physiological mechanism, a holistic approach is a natural over-compensation; but it is not practical strategy. No team of experimenters can tackle all aspects of brain physiology at once. Norbert Wiener has very properly stigmatised holism as a bogey: "If a phenomenon can only be grasped as a whole and is completely unresponsive to analysis, there is no suitable material for any scientific description of it; for the whole is never at our disposal. However, it is always possible for a slovenly worker to sweep the crumbs of his ignorance behind those parts of the phenomenon which are accessible to us."

If, and insofar as, mentality must be defined as a function of the whole assembly of possible brain conditions, then, and to that extent, attempts to describe mentality in physiological terms are vain. But, as already seen, to an increasing and still surprising extent, mental phenomena *are* accessible to analysis, *can* be studied in groups of manageable size, *can* be predicted from experimental observations.

Though we cannot expect to enquire into all the phenomena all the time, technical stratagems are available which permit study of nearly all the phenomena for a short time, or of some phenomena for nearly all the time. All the while, diversionary attacks in patrol strength are kept up in hundreds of clinics and laboratories, and from their captures is being built up a system of intelligence and espionage which is reducing little by little the domain of the central enigma. We have seen that the intricacies of learning and abstraction, the perspectives of personality and imagination, the labyrinth of original fantasy, can all be explored and charted in objective records.

Some tracts of these regions are less accessible than others; advance parties have scarcely done more than establish depots for the benefit of their successors. But there is no insuperable obstacle to be seen. The operations of learning indicated that the evolution of novel and effective ideas and behaviour depends upon the continuous assessment by the brain of the likelihood that several events are significantly related. Significance is a matter of regular repetition, of patterns against chaos. Quite simple models can be seen to extract meaning from coincidence, and snatch the little that is possibly true from the much that is probably false.

Simple machines go no further than this. From a statistic one may conjecture, but should not assert, a *cause*. But the notion of causality lurks not far behind these operations. When thunder is always and only preceded by lightning, a man or a machine legitimately infers that flash means bang; to go a step further and enquire whether flash *causes* bang is a temptation to which both flesh and metal are inevitably exposed. The Swiss psychologist Piaget has shown that children do in fact construct their ideas of causality in this way, by an ingenuous development from *post* to *propter*.

Wise men, too, are apt to conjure a cause from a constant antecedent. The wisest of them include in such arguments the reservation that when two events are constantly associated they may be the joint effects of a common cause, and to this they are inclined to give a name which may be Zeus in one age or Electricity in another. The more sophisticated successors to our child-like models, as a possible consequence of their abstractive power, if they do not discover their maker will evolve their own *deus ex machina*. Such a deity may be defined as the rate of loss of detail and certainty due to repeated reflection in a system of infinite regress. Warren McCulloch has reminded us that Spinoza called consciousness an idea of ideas, and Bertrand Russell has warned us of the danger inherent in statements which are themselves members of the class to which they refer. For those who find such exercises refreshing, the generation of transcendental propositions by mechanical contrivances may prove a welcome change from merely human perplexities.

The practical consequences of physiological infiltration into philosophical domains will be only as serious as the community cares to make them. But it may be assumed that no mental theory or practice is likely to survive which does not take into account the principles of cerebral functions revealed by physiology any more than the practice of medicine can ignore other physical functions. Already the new type of psychiatrist, in touch with centres of physiological research, has adopted a new outlook. The impact has not yet reached the great conservative body of that profession, and perhaps never will do so, especially in the United States, where psychiatry spread in a sweet flood of affluence and crystallized hard. But one may look forward with certainty to the time when brain physiology will be at least as much of a prescribed subject for the psychological and medical student as general physiology is for the latter today. This will produce psychiatrists as well informed about the organ of mentality as the medical practitioner is about the organs with which he has to deal. Perhaps, later still, there will come an integration of the two sciences in a single science of health based, not merely on familiarity with pathological conditions of mind and body, but on a full understanding of wholesome life and how the errors and chances of unhealth can be mitigated, avoided or exploited. Sapience, as it was noted when first distinguishing between the human and other brains, has just that capacity: we can make our errors in a thought and reject them in another thought, leaving no trace of error in us.

Frequent efforts to integrate or combine the sundry schools and disciplines of scientific and mental theory have been made. The loop of Cybernetics joins many of the subjects by their less tattered corners and has promoted many valuable and otherwise improbable intimacies. America is a great incubator of synthetic cultures. In the latest of these, the Dianetics of Hubbard, there appears to be both a nominal and a formal solecism. The aim was apparently to unite the principles and practices of Freud, Jung, Adler, Pavlov, Behaviourism, Faith Healing, Christian Science, Autosuggestion, Yoga and Theosophy into a single practical system of analysis and treatment. The resultant is the lowest common multiple of all these cults, incorporating their crudities and exaggerations, ignoring their subtleties and implications. This is something to beware of, for what we need is to preserve and cultivate just these growing-points of science, not to arrange in arbitrary style, like cut flowers, their sterile and exotic efflorescence.

But this borderland of the unpredictable is far from the planned field of education. Here the future can be more reasonably discussed, for the profession is more reasonable than any other, miracles not being demanded of the teacher as of priest or medicine-man. Patiently throughout the ages he has watched the input and output of generations of Black Boxes, varying the input according to results by guess and by God. To have brought so many millions to such a promising level of mental training, and with so few casualties, is an achievement beyond the success of any other profession. In these days, however, the casualties are increasing, as every teacher knows. There are those who are put out of action— the defectives and delinquents who will never return to normal school life. There are the wounded—children who may be the brightest and boldest of the class, whose feelings are hurt and frustrated by its restrictions. And there are the missing, the few already left behind, the many stragglers still vainly shepherded.

An instance has been given of the manner in which brain physiology can be useful in the understanding of delinquent children. By the same kind of approach that led, through location of tumours and other clinical aids, to better knowledge of the human brain, the study of normal childhood is now being reached. It promises in particular assistance in the problems of unhappy, backward and brilliant children, problems that multiply as the integrity of the family is corrupted and pressure on the school system increases. Just when teachers are becoming more aware than ever before that nearly every child, and especially every unusual child, would benefit from individual attention, they are being asked to take larger and larger classes. They know that, try as they may, even when the mind is most tender, they cannot mould them all alike; and physiology endorses this result of their Black Box observations. Only in a totalitarian country would that be possible, using its appropriate shallow mould. Doubtless for every Nazi fanatic was sacrificed an anonymous hecatomb. Moscow does not publish casualty lists, but we know enough about the working brain to predict that a scheme of education based on misinterpretation of Pavlov and denial of Mendel is liable to produce mental disorder in proportion with the imbecility of the premises. For this implies that totalitarian education must be based in principle upon trying to condition a fault in the conditioning mechanism of the brain, and a fault in this, it was found, is the one physiological activity that cannot be conditioned. You may learn nothing, and get away with it; but you cannot in sanity learn not to learn. The mechanism breaks down, sooner or later, when these natural functions are tampered with; the mind is flattened into a shallow mould; anything can mean anything and untruth be truth.

Western education, compounding cultural individualism and Christian egalitarianism, does not deny heredity and the recognition of types, but is in serious difficulty about reconciling political equality with a classification of types. The English system, of which selective breeding was the base and personal tutoring the pinnacle, has broken with that preferential tradition and now, in the name of equality of opportunity, attempts to separate sheep and goats at the early age of eleven. Clerical examination is notoriously fallacious in inverse proportion with the age of the subject, and if we are, anyhow, to create new castes it would obviously be wiser to have more than two. Here, meanwhile, the assistance of brain physiology can be offered with assurance in present conditions and as a help in bettering them. In the pressing problem of the backward or unusual child, take this case, for example: A boy was sent for EEG examination because he was so backward that there was anxiety about the state of his mind; at the age of nine he could not read. How should he be classified? The results of intelligence tests unequivocally classed him as defective, since he could not even read the questions. His brain record revealed nothing more ominous than large persistent alpha rhythms, suggesting that he was a pronounced P-type, like Peggy in our fable. That was all. Being a persistent non-visualist, it was useless trying to teach him to read pictorially. Properly taught, as an abstract thinker, he may turn out to be a Senior Wrangler.

Individual education aimed at the development of understanding and expression in selected pupils; mass education has a more limited goal—the training and tuning of all as selective receivers. To an increasing extent the measure of literacy is becoming the ability to read instructions rather than the power to report observations or express opinions. The cynical justification for this evolution is that only a minority can be expected to observe anything worth reporting or to have any opinions worth expressing. Experience and statistics may support this contention, but if its application is not to stultify pupils and teachers alike, means must be found to extricate from the button-mould the exceptions needed to prove the rule, to test any such generalisation to breaking point.

The whole system of civilised life depends, as we all know, on communication; but, unlike the channels of the brain, it is becoming more and more a one-way communication, from the top and centre down and out to the inert receivers who, even in the best conditions, can respond only by a unitary vote. It has been suggested that the greater an animal's brain the more its survival depends on the nature of its play. Human society devotes an enormous proportion of time and energy to play. In Western countries the invention of concerted music, team games and family holidays is recent, and already, it seems, precarious. Perhaps the most ominous feature of mechanised civilisation is that the ludicrous devices demanded for entertainment do not lend themselves to two-way operation.

Society seemed for a time to reflect in its diversity, plasticity and adaptability the generosity of brain function; now it seems to be degenerating into something more like a spinal cord, able to receive instructions and implement reflex co-ordination but incapable of initiating any independent or original idea. A passive solitary child gazing at the screen of a television receiver amuses only itself—the need to gaze does not promote or evoke habits of creativeness or generosity. True, to those who complain that things are not what they used to be, there is always the reply that they never were. What we fear from mechanical entertainment may well have been foretold as a consequence of writing and printing, of gramophones and radio. But have not these forebodings been to a great extent confirmed? The more artistic expression comes to depend on special technical devices, the more restricted is participation in it. Habits of play that depend upon mechanised art and professional sport have no need for private dreams and make-believe. H. L. Mencken once remarked that the more a man dreams the less he believes; so perhaps the less a child's make-believe, the more readily he will later grasp at ready-made beliefs. For Alice in Movieland the future looks drab; Tom Sawyer will have few adventures at the television set.

The notion of a past Golden Age of individuality and enterprise, however, has but flimsy foundations; for most, those times were dross, or leaden oppression. Our libraries contain the cream of the past; naturally we find the whole milk of the present thin in comparison and the processed cheese of the future an indigestible substitute.

In what ways, then, can knowledge of brain physiology help to predict and control the course of history? The tentative classification of simple brain records has indicated a wide prospect of diverse and fluctuating personalities. Intelligence, as estimated by arbitrary and already obsolescent tests, finds no parallel in our tracings; but versatility, ductility, and certain special imaginative aptitudes, are beginning to be recognised as dynamic interrelations and transformations within the framework of normal variation. Note that the classes so adumbrated are not based on what a person can or cannot do, but on the *ways* of doing things. Further, nothing suggests that one type or one way is better than another. The voice of the physiologist is scarcely more than a whisper; but, if it were called upon to testify, it would derive equality of responsibility from diversity of endowment. To each, it would murmur, according to his capacity, from each according to his needs.

The factors we seem to recognise, and more particularly the names we give them, may be superseded as understanding widens, but whatever changes further experience may dictate the picture will not be any less variegated; and it is in helping to select the more exceptional and less hardy strains for culture that the brain physiologist may find the most exhilarating challenge. The late Dr. Montessori initiated her revolution in free discipline and sensory training by raising the attainment of mental defectives to the level of the State examinations in Italy at the turn of the century. The physiology of the brain suggests that the success of her empirical methods was due to the simultaneous exploitation of the individual characters and the common properties of her pupils' brains.

Amplified by understanding of the basic functions involved, the physiological training of unusual brains may have results that are quite unforeseeable. We are so accustomed to mediocrity as the arithmetic mean of our neighbours' limitations, that we can scarcely conceive of the intellectual power of a brain at full throttle. Natural geniuses are personally known to few, and their capacity can be estimated only by their peers. Most of them confess to disabilities which would be fatal in most circumstances. Of the 1 per cent of our population that we should expect to excel, only a small proportion reach years of discretion undistorted by their upbringing. The 10 per cent that we class as defective but capable of education, are as often as not tormented by competing outside their class or type and add the neurosis of inferiority to the modesty of their understanding.

In those few effective outstanding personalities whose brain functions we have been privileged to study, versatility of both function and brain activity seems the only common factor. In most people an analysis of thirty seconds of brain activity is enough to provide a representative sample; in that time, left to itself, the brain goes through its modest repertory of motions. But in the efficient genius brain, the motto seems to be *semper aliquid novi;* several minutes of analysis are needed before the picture begins to repeat itself, even in the most tranquil conditions. We are beginning to have some insight into what Margery Fry has called the "climate of delinquency," and education establishes a seed-bed for gifted specialists. It is surely time that we studied also the conditions which favour the development of the versatile genius.

Do such suggestions imply or depend upon the predominance of hereditary influences? So far, it is true, the study of brain physiology has tended to emphasise the importance of genetic factors. Lennox, who was one of the first to explore the electrical accompaniments of epilepsy, even went so far as to assert that "the electroencephalogram is an hereditary trait," and to urge that an EEG be taken of every citizen as a condition for granting a marriage licence. This bias in favour of nature as opposed to nurture may be justified by further experiment, but so far no crucial observations have been reported; even those on identical twins are inconclusive.

At the present time this assumption of genetic prepotence is a hindrance to the development of anything like a physiological sociology. Environmental influences are so much easier to criticise and control; they provide a safe battleground for the social reformer and the revolutionary. The popular view is that, apart from a few hard cases, the majority of unhappy or anti-social people are the victims of circumstance and can be restored to conformity by circumstantial adjustment. This assumption is based on the hopeful and kindly ethic of Western civilisation, derived from Christian doctrine and enshrined in statute law as in everyday habit. One may describe it as the "ewe-lamb" hypothesis as opposed to the less attractive "black-sheep" theory.

It is much easier to classify influences than to understand them. The crude division of all human attributes into "inherited" and "acquired" is excusable but quite unreasonable. Even in the simple models of behaviour we have described, it is often quite impossible to decide whether what the model is doing is the result of its design or of its experience. Such a categorisation is in fact meaningless when use influences design, and design use. We can infer the existence in human nature of intricate inter-action between inherited and acquired characters, if only from the fervour with which adherents of both causes deny the truth of their opponent's testimony. In such a system, when variables are subject to mutual influence, it is characteristic that the denials of both sides should turn out to be false and their assertions true; for quite superficial observation reveals that almost anything *may* happen, whereas the most painstaking analysis is necessary for anyone to define what is impossible.

A reasonable question for us to ask is: what chance is there of the human brain evolving further? As far as is known today, there are two ways in which evolution can take place. First, by the occurrence of major mutations or sports important enough to increase the chances of survival of the individual and to produce a new species. Second, by selection and hybridisation of minor mutations to produce a new variety. The fundamental difference between the two mechanisms is that, while the first initiates a species with a different gene structure, which can no longer breed with the parent stock, the second produces only slowly a new variety which can still breed with the original.

All human beings are one species, inter-fertile and mutually attractive, as the great variety of half-breeds testifies. The differences between ethnic groups are mainly in hair and hide, and since the brain is developed from embryonic tissues similar to those which give rise to skin and hair, we might expect some difference between the brains of, say, Caucasian, Mongol and African. Slight differences there do seem to be, as Mundy-Castle has found in Africa, but scarcely more than can be accounted for by the varying traditions and standards of life and nourishment. And did not the brains of these people establish and elaborate these traditions, accept or modify these standards? Unluckily for our peace of mind, there do exist precipitous gradients of economic and political differences; but brain physiology detects no incompatibilities; indeed in regions where the shades and textures of the human race are freely blended, there is found the richest variety, the most lively growth.

The appearance of a new human species with an even larger and more adaptable brain seems extremely unlikely, short of major catastrophe. Such an offspring would be sterile unless it met a similarly improbable mate. The fate of a sport or monster of this type, moreover, in relatively peaceful conditions would doubtless be to languish in a colony of mental defectives or lunatics. The super-brain would almost certainly need a superior and protracted education. Arithmetical prodigies only prove the rule; their specialised talent, unlike the versatility of genius, seems to need little education. From pre-ape to man, the period of mental adolescence has increased tenfold; to exploit a similar evolutionary jump now toward after-man would probably entail a century of super-schooling. For dramatic brain evolution to have any chance of survival, more than a swollen head would be needed; we should have to consider not only Man and Superman but Back to Methuselah. The whole anatomy and physiology of our bodies would have to be modified to maintain life for millennia and support not merely the weight of a heavier head but the accumulating stress and burden of encyclopaedic wisdom perpetually displayed. A centenarian schoolboy is a freak few of us would welcome and only a planetary cataclysm could foster such a monster.

Minor changes over a great span of years are no doubt proceeding continuously; it is to be assumed that the brain which can influence environment will continue to be adaptable to changes of environment. Two particulars of its present limitations have been mentioned earlier. One of them was the exceptional sensory relation between touch and brain; the puritan movement and modern sophistication are protests against this vestigial enthrallment, and may well come to have survival value in a rational world. The other particular was the limitation of speed imposed on our sensory and mental reactions by the rate of the alpha rhythms. A faster rhythm has unquestionably survival value in a way of life calling for ever speedier decisions and actions. The action difference between an alpha rhythm of 8 c/s and one of 13 c/s can be estimated, for instance, in the emergency of stopping a car: at 50 miles an hour a driver with the faster rhythm would stop his car 5 feet short of the point reached by a driver with the slower rhythm. Similarly, pedestrians and cyclists with the faster alpha rhythms have a better chance of escape.

The chances of a major mutation being advantageous are rarely better than one in a million, and in normal circumstances the human mutation rate is very slow. Conceivably, if a populous, fertile and temperate part of the planet were isolated from the rest of an atomised world, and for a time subjected to considerable gamma radiation, viable sports with superior brains might survive long enough to propagate their kind, supplant their ephemeral forebears, and later populate this or other worlds. Such events have been imagined often enough in fiction and we may suppose that anything we can imagine is remotely possible; indeed in the very remote future changes of this order are inevitable, and may already have happened in some other place.

Our imagination and perhaps our hopes may stray toward the people of the stars, but our concern is not with these. For us, and any that we may call ours, the future is more placid; yet more arduous. The foreseeable future of brain is more a matter of hard study and design. We need not yearn for greater masses of grey matter—we already dispose of enough nerve units to enumerate in their permutations every particle of Eddington's universe. More probably the next chapters of brain history will be not so unlike the first; the theme has been, is, and probably will be, delegation without specialisation.

During the last two generations the rate of accumulation of knowledge has been so colossally accelerated that not even the most noble and most tranquil brain can now store and consider even a thousandth part of it. Humanity is in a critical state compared with which the constitution of the Dinosaur and the community of the Tower of Babel were triumphant. Our first response to the challenge of this deluge has been a tactical success, though it contain the seeds of strategic disaster—specialisation without delegation, the inversion of pragmatic justice. The professor in his lair can always find an expert or an abstract to patch the gaps that inevitably yawn in his knowledge as his subject swells; but, to provide his auxiliaries, other professors must train the experts to write the abstracts—and bewilderment mounts in rapidly widening spirals. The economics of information has its Gresham's Law, too—half-truths drive out full understanding.

Continuation of the sectarian process of specialisation could only lead to one result, the creation of an irresponsible scientific priesthood, preoccupied entirely with its liturgy and its mysteries; and, in due course, to a popular revulsion from scientific knowledge and a slump of scientific credit that would usher in a dark age as vicious and prolonged as the aftermath of an atomic war.

The root of this evil is that facts accumulate at a far higher rate than does the understanding of them. Rational thought depends literally on ratio, on the proportions and relations between things. As facts are collected, the number of possible relations between them increases at an enormous rate. When all but a few relations can be excluded as "impossible," the subject they relate to is called easy; but such confident exclusion has so often proved wrong in the past that scientific workers are loth to deny possibilities. The remedy would seem to be to avoid burdening human brains with mechanical tasks. As Wiener has passionately declared, to do this is to stultify science and corrupt human relations. There are already plans for instruments to end this bondage of the brain, for delegating to machinery the menial but essential tasks of fact arrangement and appraisal, as the management of our bodies long since was delegated. Even quite simple equipment will give judgments at least as good as our own on the validity of our hypotheses, and suggestions as to how we can improve them and plan better experiments.

Such machines have been decried and ridiculed, prostituted to the computation of mass destruction, and fictionalised into superhuman idiocy and malevolence. As a child frightened by a teased puppy will say he met a bear, so we tend to project into these docile slaves of the laboratory our feelings of guilt, apprehension, inferiority and insignificance. In fact they are domestic servants as truly the friend of man as are the dogs and horses he has fashioned from the raw material of animal species. Some of the more modest gear, designed specially for the study of brain functions, has been described here; in all such equipment there is the shadow of a brain to which can be delegated some mechanical duties—otherwise we should take a lifetime to sort out the data from a week's work.

There is also more personal co-operation. The incursion of higher mathematics and new algebras into what we thought was a quiet experimental science is horrifying to those of us whose education in these arts was neglected. However, we have powerful allies, mathematicians and engineers of high standing. Wiener, Pitts, Mackay, Shannon, Weaver, Rashevsky and his school, have all blazed trails into the new territory. The paralysis or revulsion which assails most of us when a mathematical concept flashes our from behind our blinkers is allayed by the efforts of such masters of popular exposition as Bronowski and Hoyle, who, appreciating the strangeness and novelty of such notions, insinuate most gently into their discourses the relation between the basic abstractions of mathematics and the material entities of the living brain. The schoolroom bogeys of irrational and imaginary numbers need no longer intimidate those who have *seen* a brain manipulating incoming signals to extract from them a notion of number and meaning.

The extreme specialisation and segregation of the present epoch is a novelty in human affairs. If left unmitigated it could spell ruin. But the remedy is traditional in principle, an extension of custom rather than a revolution. Libraries, almost as old as the history of the race, are, in a sense, the racial memory. Their importance and power may be gauged by the hostility they have evoked and the fury with which they have been attacked by established or upstart authority; the Burning of Books has been a standard sacrifice of the tyrant from Alexandria to Berlin. Libraries solved for many generations the problem of laborious copying by hand; printing dealt as surely with the re-birth of curiosity. For each emergency the brain of man invents a remedy, a technique of delegation, so as to maintain intact its unspecialised competence.

In the present crisis, which threatens that faculty more seriously than ever it was threatened in the past, something more than an overgrown racial memory is needed; no mere library of facts and opinions will constrain and direct the explosive force of scientific exploration. The machines that flash and click in our laboratories now are the first forms of the living brain's extended life, the rudiments of racial understanding, as Gutenberg's first printing presses were the forerunners of the Reformation.

The storage and arrangement of facts is familiar to us, whether in print or on film or on cards; but the mechanisation of understanding, the use of machines to determine significance, abstract meaning, and conduct logical arguments, is as strange and repulsive to many people as the notion of moveable type was five centuries ago. Yet we have seen that in their simpler forms these intellectual processes can be analysed and imitated without great difficulty.

The exteriorisation of tedious or controversial reasoning will no doubt have as profound effect upon the brain and society as the introduction of skilled and respectful servants has on a humble household. No man, we may suppose, will be a genius to his own machine, and mediocre thinkers will be as hard hit as imperfect scribes were by the printing press. But the future of the brain is more intriguing than a mere holiday from drudgery, for it is only when the servants of thought have done their work and retired unobtrusively to their quarters that the master brain can discover its own place and settle down at last to its proper work.

That is a reflection of the laboratory, but it is not the end of the matter. The proper study of mankind, if we are creatures of purpose, has its proper application. Enough is known today about the living brain to reduce material waste and human misery—in education, in correction and in the development of mature personal relations. In an epoch of rapidly expanding human power, the application of this new and growing self-knowledge will concern not only the domestic or national future but the destiny of the human species.

APPENDIX A

An Electric Model of Nerve

In GENERAL, it is legitimate to study a model of a mysterious process if three conditions are fulfilled: 1. Several features of the mystery must be known. 2. The model must contain the absolute minimum of working parts to reproduce the known features. 3. The model must reproduce other features, either as predictions, or as unexpected combinations.

There are several legitimate models of nerve. The earliest ones were simple circuits containing resistance and capacitance, and copied only the passive properties of nerve—they did not propagate an impulse or even suggest how an impulse would be propagated. They had the advantage of drawing attention to the similarity of a nerve to a leaky cable such as a submarine telegraph line. Since the mathematical equations relating to leaky cables were worked out during the last century, physiologists could apply rigorous and well-tested notions to those passive features of nerve which the "leaky cable" models reproduced. Later, electro-chemical models were discovered. The best known of these is, incongruously enough, an iron wire in strong nitric acid. The acid forms an oxide film on the wire so that the iron within does not dissolve. This film is "passive" but breaks down when scratched or stimulated electrically, for example, when the wire is touched with a piece of zinc. When stimulated, an impulse passes quite quickly down the wire, and this impulse has

THREE ELEMENTS OF MODEL TO SHOW EXCITATION AND PROPAGATION

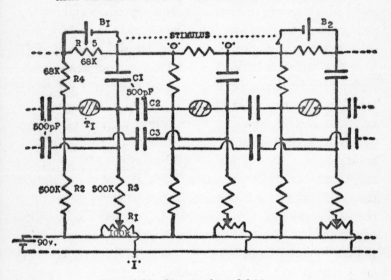

Figure 21. Circuit of Model Nerve.

many of the properties of a nerve impulse: it is a vortex ring of electro-chemical action. During the passage of the impulse the passive film is decomposed momentarily, and the nitric acid attacks the iron with the evolution of nitric oxide. A fresh passive film is formed and this is "refractory" for a short time; the wire cannot propagate another impulse immediately after one has passed by. This is a good dynamic model but has the disadvantage that the nature of the passive film is almost as mysterious as the nerve fibre itself; it is not very satisfactory to equate two unknowns.

It is possible to retain the simplicity of the leaky cable models and add to them a dynamic element to represent the mechanism in a nerve which provides the miniature electrochemical explosion seen as an impulse. The circuit of a working model is shown in Figure 21. The capacitors and resistors provide the elements of a leaky cable, and the battery maintains a steady voltage such that the "inside" of the model is negative to the outside. The addition to the circuit which endows it with the power to propagate an impulse is the neon tube, also connected, in effect, between the inside and outside and biassed to a few volts below its striking voltage, which in the case of the miniature tubes used for the embodiment of this circuit is between 50 and 60 volts. Every element (consisting of resistors R_1, R_2, R_3, R_4, the neon tube and the capacitor C) is connected to the adjacent elements on both sides through the capacitors C_2, C_3 and C_4, C_5. These capacitors join points of opposite polarity of the neon tubes, so that they may be envisaged as being in the form of a crisscross connection, a sort of lazy-tongs arrangement extending down the chain of elements.

Providing all the neon tubes are below their striking voltage, the system is stable and inert. If, however, a voltage is applied as indicated by the external battery B_1 or B_2, the voltage across one of the neon tubes rises, and when it reaches the striking threshold the tube ionises and partially discharges the capacitor C_1. When the tube ionises the voltage across it drops to the extinction level. This voltage drop is applied to adjacent tubes through C_2, C_3 and C_4, C_5, in such a sense as to *increase* the voltage across them (owing to the criss-cross connection) and they accordingly strike in their turn. The impulse is thus propagated to both ends of the chain at a velocity depending on the values of the capacitors and resistors.

In the working model which has the values shown in the figure the conduction time is about 0.2 m. sec. for an element. The action potential has a duration of about 10 m. sec. —rather longer than most nerves; the absolutely refractory period is about 5 m. sec. and the relatively refractory period about 15 m. sec. Twenty such elements represent five to ten centimetres of frog motor nerve at about 10°C. The action potential, excitability, characteristics, chronaxie, accommodation, space constant, and so forth, have values which are interrelated in much the same way as in a nerve. The size and effectiveness of the impulse at any point depend only on the state of the element at that point, so that propagation is decrement-less and all-or-none.

A synapse may be formed by omitting one of the capacitors such as C_2 or C_3. If then the threshold of the element on one side of the half-break is raised by lowering the bias voltage it can be stimulated by an impulse arriving from the unchanged side and can just stimulate on the other side through the single capacitor; it cannot, however, be stimulated through the single capacitor. Propagation is therefore unidirectional and is then found to be frequency-sensitive. Very low and very high rates of stimulation are relatively ineffective and there is a fairly marked optimum range of stimulation rate. This implies facilitation; two impulses may be effective when one is not. A junction between several such chains exhibits all the properties of simple reflex systems, such as recruitment and occlusion.

Inhibition may also be demonstrated. Wedensky inhibition is seen when the threshold of one element in a continuous chain is raised. A low-frequency train of impulses will pass this depressed element, but a high-frequency one will not. An inhibitory synapse is made by connecting a single capacitor from the positive side of one element to the positive side of another. A train of impulses in the "pre-synaptic" stretch then inhibits a spontaneous discharge on the other side of the inhibitory link; the degree of inhibition is a function of the frequency of the impulses in the pre-synaptic stretch.

Spontaneous activity is easily produced by raising the bias voltage of one tube so that it regularly strikes and discharges its condenser, which then charges up again, and so forth, as in the simplest relaxation oscillator circuit. The top limit of spontaneous discharge frequency is fixed by the absolutely refractory period and in the actual model is about 200 impulses per second.

Altogether the model seems to display 18 or so of the known properties of nervous and synaptic excitation and propagation. Many of these properties were not foreseen as following inevitably from the elementary features originally specified as imitable. For instance, an inhibitory end-organ can be connected to one end of the chain and an inhibitory synapse arranged in the middle of it. This is achieved by connecting the anode of a photo-cell through a capacitor to the positive side of the first element of the pre-synaptic chain. With this arrangement, when light falls on the photo-cell its anode becomes more negative and transiently decreases the excitability of the element to which it is connected. The adaptation rate is governed by the time constant of the coupling capacitor and resistors. Now, if the post-synaptic stretch is set to discharge at a moderate rate, and the pre-synaptic stretch at a high rate, the activity of the latter will partially inhibit that of the former. When a light is shone on the photo-cell, the pre-synaptic discharge will be inhibited, but in consequence the post-synaptic activity will be *dis*-inhibited. Further, if in identical conditions the photo-cell is illuminated by a *flickering* light, at a moderate rate of flicker, inhibition at the end-organ is converted into excitation; the system responds with a rebound at the *end* of each flash. Accordingly, with flicker, the discharge rate in the pre-synaptic chain is augmented instead of being reduced by steady light, and the post-synaptic activity is still further inhibited instead of being dis-inhibited.

Effects such as this are in fact seen in the central nervous system; a change in stimulus frequency has often been found to invert the response, and the anomalous effects of flicker have been described in some detail. It may well be that these otherwise rather puzzling phenomena may be explicable in terms of the peculiar properties of rapidly adapting inhibitory synapses, displayed so clearly in this simple model. The fact that this model is affected by and produces electrical rather than chemical or mechanical changes should be regarded as a convenience and a coincidence. It is not, of course, proof that the electrical changes in nerve are the essence of nervous action. The model is simply the analogue of one set of familiar mathematical expressions relating to passive networks linked by a non-linear operator in the form of a discharge tube. It could quite well be formed of chemical or mechanical parts and does not in theory contain more information than do the algebraic equations. Its advantage is that, being a real object, it has constant dimensions; hence its predictions are more explicit and detailed than those of the equations, in which the constants are rather more arbitrary and independent.

APPENDIX B

The Design of
M. *speculatrix*

As ALREADY emphasised, the simpler a model is, the more helpful it is likely to be. This is a model of elementary reflex behaviour and contains only two functional elements: two receptors, two nerve cells, two effectors. The first receptor is a photo-electric cell, mounted on the spindle of the steering column so that it always faces in the same direction as the single front driving wheel, which is one effector. In the dark the steering is continuously rotated by the steering motor, the other effector, so that the photo-cell scans steadily. The scanning rotation is stopped when moderate light enters the photo-cell, but starts again at half speed when the light intensity is greater—the dazzle state. The driving motor operates at half speed when scanning in the dark, at full speed in moderate or intense light. The other receptor is a ring-and-stick limit switch attached to the shell, which is rubber-suspended. When the shell touches something or when a gradient is encountered, it is displaced and closes the limit switch. This connects the output of the "central nervous" amplifier back to its input through a capacitor so that it is turned into a multivibrator. The oscillations produced by the multivibrator stop the circuit from acting as an amplifier so that simple sensitivity to light is lost; instead, the connections alternate between the "dark" and "dazzle" states. The steering-scanning motor is alternately on full- and half-power and the driving motor at the same time on half- and full-power. The effect of this is to produce a turn-and-push manoeuvre. The time-constant of the feedback circuit is selected to give about one third of the time on "steer-hard-push-gently" and two thirds "push-hard-steer-gently." This seems to give a prompt response to the first contact with an obstacle, a reasonable chance of getting away, or getting through a gap—and a short after-discharge to ensure final escape. Though there is no direct attraction to light in the obstacle-avoiding state, the feedback time-constant is shorter when the photo-cell is illuminated, so that when an obstacle is met in the dark the avoidance drill is done in a leisurely fashion, but when there is an attractive light nearby the movements are more hasty.

The electrical circuit is shown in Figure 22. This is only one of many possible arrangements, but is probably the simplest in components and wiring. The photo-cell is a gas-filled type, and generally needs no optical system, a single light-louvre giving sufficient directionality. It is convenient to connect the tube between the grid of the first amplifier tube and the negative side of the 6-volt accumulator needed to run the motors. This gives sufficient sensitivity for an ordinary flash-lamp to control the device, and avoids overdriving the minia-ture tubes; but other connections are possible, particularly if other types of amplifiers are used. The grid of the input tube is connected to the positive side of the 6-volt battery through a 10 meg. resistor; illumination of the photo-cell can there-fore only change the bias on the input tube from zero to about 4 v. negative. In the dark the first tube, having zero bias, passes its full current, and the relay in its anode is "on." This tube is a triode or a triode connected pentode. The relay should have a resistance of about 10,000 ohms, or rather less than the anode impedance of the tube, and a single pole change-over contact. The resistance of this relay and the

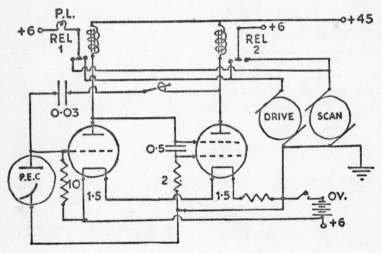

Figure 22. Circuit of *M. speculatrix.*

anode impedance of the first tube form a potentiometer which fixes the screen voltage of the second tube. The anode of the first tube is thus connected directly to the screen of the second and also through a 0.5 mf capacitor to its grid. This provides a relatively high gain for *changes* in illumination and steady-state amplification when the input is larger. Living systems commonly differentiate or "adapt" to moderate changes in stimulus strength but respond steadily to intense stimuli. The effect of this coupling is to permit transient interruption of the scanning motion when a faint light enters the photo-cell, thus gradually bringing the model on to the beam at a distance, then a steady inhibition of scanning when the light is brighter or nearer. The relay in the anode of the second tube is of the same type as the first, but the moving contact goes straight to the positive terminal of the 6-volt battery instead of through the pilot light. The stationary contacts are connected in the same way, "on" to the driving motor, "off" to the scanning motor, in both relays; in faint light, relay 2 is closed momentarily; in moderate light it is held closed; and in bright light it remains closed but relay 1 opens, thus providing for swerving away from a bright light.

The pilot light, which is in series with the moving contact of relay 1, is short-circuited when relay 2 closes, and is therefore extinguished when the driving motor is turned to full power and the scanning movement is arrested by light. When the light from the pilot bulb is reflected from a mirror into the photo-cell, it is turned off, but the disappearance of this light restores relay 2 to "off" and the light appears again. This provides the self- and mutual-recognition modes. Several circuit variations are possible; the effect of light can be reversed by inverting the photo-cell connections, and the relays may at the same time, or independently, be reversed also. The photo-cell may be positioned to face in the opposite direction from the movement, and so forth.

The mechanical design is usually more of a problem than the electrical. Remembering that all power must be carried, a clear optimum exists for battery capacity, chassis dimensions, motor power and the like. There is not a great choice of motors; those used for driving small home-constructed models are adequate but not efficient owing to their disproportionate frictional losses. It is often advisable to re-bush the bearings of these motors with a self-lubricating bearing. The gear trains to the driving and scanning shafts are the most awkward parts for an amateur constructor. The first model of this species was furnished with pinions from old clocks and gas-meters. The ratio should be about 100:1 for the driving and 180:1 for the steering mechanism, and is best obtained with simple single pinions, since small worm reduction-gears are wasteful and liable to seize when forced backwards, as happens sometimes when the model is in difficulties.

The weight-distribution is critical, since the front wheel is turning almost continually through 360° even when there is no forward motion. The forward weight should be only just enough to ensure an adequate grip for the front tyre, and the centre of gravity should be just forward of the line joining the rear wheels. The front tyre should be rubber, but thin and fairly hard, so that it can turn easily. The rear tyres should be soft and larger. The wheelbase determines the turning circle and should be about 6 inches, the rear track being about 7 inches. Since the motion of the model is a combination of linear and circular displacements, its gait is cycloidal, and once it has lost a target, a complete rotation of the steering scanning shaft may be necessary to pick it up again. The continuous rotation may be replaced with a reciprocating system using a windscreen-wiper motor geared up to 2:1 to give 360° sweep; but this also has disadvantages. The model may be made into a better "self-directing missile" by using two photo-cells in the usual way, but it will be a worse "animal," for though it will keep more closely to its beam it will have to be aimed roughly in the right direction and will not "speculate"—that is, spy out the land—nor will it solve Buridan's dilemma.

APPENDIX C

A Conditioned Reflex Analogue

THE CIRCUIT in Figure 23 seems to be the simplest to display the necessary properties. It was worked out to fit on to *M. speculatrix*, but it is difficult to adjust on a moving model and should be set up on the bench first.

There are four tubes; the first, a sharp cut-off pentode, mixes and integrates the specific and neutral signals; the second, a miniature neon-tube, strikes when the voltage across the condenser in the anode of the first tube reaches its threshold; the third receives a transient from the neon-tube and maintains a damped oscillation in the phase-shifting network; the fourth is a gating heptode which receives signals both from the oscillating circuit of the third tube and from the source of neutral signals, conducting only when both are present.

The values of the components depend on the range of responses desired. The time-constant of the high-pass network C_1 R_1 determines the differentiation of the specific stimulus applied to the grid of the pentode, which is biassed back to cut off through R_1. The low-pass network C_2 R_2 controls the rate of rise and extension of the neutral signal reaching the pentode screen, which is at cut-off voltage until a signal appears.

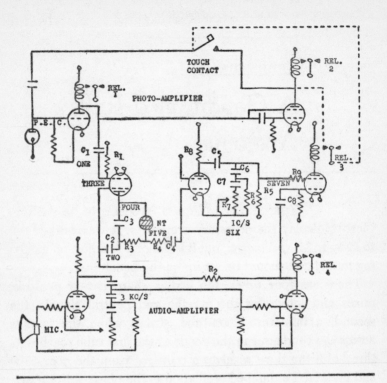

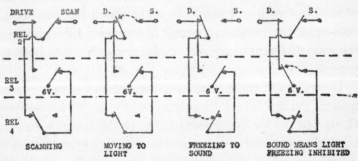

Figure 23. Circuits of *M. docilis.* (a) Functional circuit. The numbers ONE to SEVEN refer to the operations performed by each element as detailed in Chap. 7 and correspond to those in Fig. 17.

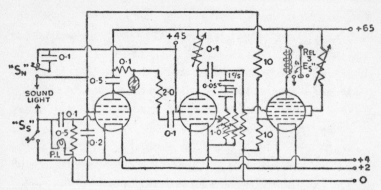

Figure 23 (continued). (b) Conditioned Reflex Analogue. Circuit details of CORA as demonstration model.

The first tube should not conduct unless both grid and screen receive positive signals and the screen signal precedes or coincides with the grid signal. If the grid signal is more than about 0.1 sec. before the screen one, the grid will have returned to cut-off level before the screen becomes positive. The bias conditions and time-constants must be so adjusted that the tube conducts only when the two sets of signals have these relations. The value of the capacitor C_3 determines the time-constant of long-term integration by which is computed the over-all time-area of coincidence between successive appearances of neutral and specific signals.

At each combination of the two signals in the effective time-order, the voltage across C_3 rises by an amount dependent on the degree of overlap between the differentiated specific and extended neutral signals. In the period between signals—which may be relatively very long—the voltage tends to fall slowly as the capacitor discharges. Stray and internal leakage resistances are usually adequate to ensure that some discharge occurs, equivalent to the gradual lapsing of unconscious memory of a coincidence. A leakage resistance of 1,000 megohms per microfarad will give a constant of 1,000 seconds—that is, about 20 minutes—by which time the voltage will fall to 37% of its original value. This is appropriate and compares roughly with the time constant of decay in some animals; with a good modern capacitor the decay time may be ten times as long.

The rising voltage across C_3 is applied continuously to the neon-tube, which strikes at some voltage between 50 and 65 volts. (The tube may be visible from outside and labelled "Insight.") In consequence the capacitor is discharged to about 40 volts, thus retaining a trace of the previous experience; and the discharge current, passing through R_3, provides a transient voltage through R_4 and C_4 to the grid of the pentode, which is also of the sharp cut-off type. Between the anode and grid of this pentode is connected a 3-stage high-pass network—$C_5 R_5$, $C_6 R_6$, $C_7 R_7$—providing a phase shift of 180° at some convenient frequency.

The amount of feedback in this circuit is controlled by the potentiometer R_7. (The point of zero feedback may be labelled "Amnesia.") The amount of feedback determines the Q and decay time of this circuit. When the gain of the tube is about 30, spontaneous oscillation will build up, and the control should be set so that the circuit is just quiescent, with a long die-away. The value of the anode resistor R_8 and the screen voltage of this tube must be adjusted carefully so that the working conditions are optimal, in order to obtain the longest possible decay time compatible with stability. (The anode resistor may be variable and used as a fine-control of Q, or "Reminiscence.")

The time, T, to decay to $1/e$ of the original amplitude is

given approximately by $Q/\pi f$; this means that the lower the frequency, the longer the decay time can be for a given Q and stability. With careful adjustment, a Q of several thousand can be attained, so that at one c/s a decay time of ten or fifteen minutes can be expected, and the rate of decline after the 1/e time is of course slower and slower.

A milliammeter in the anode circuit indicates the presence of a "memory." This meter can be labelled "Reminiscence."

The declining oscillatory signal from this circuit is applied to the signal grid, G_8, of the heptode tube through the resistor R_9, this electrode being biassed to cut off through R_5. The first grid of this tube is connected through C_8 to the source of the neutral signals, and, when driven positive at the same time as the oscillatory signals are driving G_8 positive, the tube conducts and operates the relay in its anode. This relay is arranged to perform whatever function was associated with the generation of the specific signal.

The choice of tubes and values in the original CORA was limited to those which could be obtained in battery miniature tube types. A mains-operated demonstration model is very much easier to set up and maintain. For instance, an unbypassed cathode resistor of about 300 ohms in the resonating pentode circuit provides negative feedback which makes the system more stable at very high Q values. The final mixing gate can be a double triode, but the heptode arrangement ensures that if and when the neutral signal arrives at the gate tube, and drives G_1 positive, this tube provides an additional resistance in parallel with R_5 and tends slightly to damp the oscillation in the preservation circuit—that is, the conditioned "memory" is attenuated by persistent repetition of the neutral signal alone. The effect of the neutral signal, after

conditioning is established, is only transient owing to the
decay of voltage across C_s. This temporary extinction does
not prevent the reappearance of the conditioned response
when another trial is made after a few moments.

The very high Q of the resonating circuit makes the system
extremely sensitive to small rhythmic transients repeated at
the frequency to which the circuit is tuned. This means that
rhythmic operations, such as applying the specific or neutral
signals at the frequency of the tuned circuit, may establish or
evoke an irrelevant "memory" if they change the voltage of
any of the supply circuits by as little as one part in a thousand. The supplies must therefore be obtained either from
low-resistance batteries or a well-stabilised power-pack. Separate supplies for the resonator are easy to arrange and give
least trouble.

When the model is designed simply as a bench demonstration, operation of the keys providing the neutral and specific
signals can be arranged to illuminate suitable signs to indicate the conditions being imitated; if it is desired to attach
the circuit to a mobile model such as *M. speculatrix*, the appropriate connections are as indicated. The chief difficulty
here is ensuring adequate homeostasis; when the driving motors are heavily loaded they lower the voltage of the battery
and disturb the equanimity of the learning circuit. It is
usually necessary to contrive a "circle of Willis"—that is, a
separate heater battery—in order to maintain constant supplies in spite of the demands of the motors. This feature, and
the sensitivity to rhythmic stimuli, would be serious faults in
a tool-machine but are virtues in a lifelike toy, for they exist
in animals.

When a defensive reflex is arranged, the connections are

as shown dotted; and the inhibitory links necessary to suppress a reflex incompatible with the conditioned response are also shown in the diagram of the relay connections.

The sound amplifier is self-explanatory; sharp tuning is necessary to avoid undue sensibility to random noises.

An instructive application of the circuit as a demonstration model is to make all important controls adjustable and calibrated in appropriate units, and to induce a few students to set them all so that the behaviour of the model is as sensible as they think that of a good animal should be. The settings decided upon illuminate both the theory of learning and the temperaments of the learners.

A Short Bibliography

E. D. Adrian: *The Physical Background of Perception* (Oxford, University Press, 1947).

W. R. Ashby: *Design for a Brain* (London, Chapman & Hall, (1952).

J. Barcroft: *Features in the Architecture of Physiological Function* (Cambridge University Press, 1934).

G. von Bonin: *Essay on the Cerebral Cortex* (Springfield, Thomas, 1950).

M. A. B. Brazier: *The Electrical Activity of the Nervous System* (London, Pitman, 1951).
Bibliography of Electroencephalography, 1875–1948 (Montreal, International Federation of EEG Societies, 1950).

W. B. Cannon: *The Wisdom of the Body* (London, 1932).

K. J. W. Craik: *The Nature of Explanation* (Cambridge, University Press, 1943).

Margery Fry: *Arms of the Law* (London, Gollancz, 1951).

F. A. and E. L. Gibbs: *Atlas of Electroencephalography* (Cambridge, Mass., Addison-Wesley, 1950).

D. O. Hebb: *The Organization of Behavior* (New York, Wiley, 1949).

J. D. N. Hill and G. Parr, Editors: *Electroencephalography* (London, Macdonald, 1950).

R. L. Hubbard: *Dianetics, The Modern Science of Mental Health* (New York, Hermitage Press. London, Derrick Ridgway, 1951).

L. A. Jeffress, Editor: *Cerebral Mechanisms in Behavior* (New York, Wiley. London, Chapman & Hall, 1951).
The Hixon Symposium, consisting of 7 contributions by au-

thorities on mathematics, psychiatry, psychology and physiology.

P. DE LATIL: *La Pensée Artificielle* (Paris, Gallimand, 1952).

K. LORENZ: *King Solomon's Ring* (New York, T. Y. Crowell. London, Methuen, 1952).

J. M. D. OLMSTED: *Claude Bernard, Physiologist* (London, Cassell, 1939).

W. PENFIELD and T. RASMUSSEN: *The Cerebral Cortex of Man* (New York, Macmillan, 1950).

C. PINCHER: *Evolution. Reason Why Series* (London, Herbert Jenkins, 1950).

D. RICHTER, Editor: *Perspectives in Neuropsychiatry* (London, H. K. Lewis, 1950).

R. SCHWAB: *Electroencephalography in Clinical Practice* (Philadelphia, London, W. S. Saunders, 1951).

C. S. SHERRINGTON: *The Integrative Action of the Nervous System* (Cambridge, University Press, 1905, 1947).
Man on His Nature (Cambridge, University Press, 1940, 1951).

L. L. WHYTE, Editor: *Aspects of Form* (London, Lund Humphries, 1951).

N. WIENER: *Cybernetics* (New York, Wiley. London, Chapman & Hall. Paris, Hermann, 1948).
The Human Use of Human Beings (Boston, Houghton Mifflin, 1950).

J. Z. YOUNG: *Doubt and Certainty in Science* (Oxford, Clarendon Press, 1951).

Electroencephalography and Clinical Neurophysiology. An International Journal, published quarterly in Montreal for the International Federation of Societies for Electroencephalography and Clinical Neurophysiology.

Index